UNIQUE ENVIRONMENTALISM
A Comparative Perspective

NONPROFIT AND CIVIL SOCIETY STUDIES

An International Multidisciplinary Series

Series Editor: Helmut K. Anheier,
    London School of Economics and Political Science, London, United Kingdom and
    University of California Los Angeles, Los Angeles, California

UNIQUE ENVIRONMENTALISM
A Comparative Perspective

Gunnar Grendstad, Per Selle, and Kristin Strømsnes
University of Bergen
Bergen, Norway

and

Øystein Bortne
University of Stavanger
Stavanger, Norway

 Springer

Gunnar Grendstad
Department of Comparative Politics
University of Bergen
Bergen 5007
NORWAY
gunnar.grendstad@isp.uib.no

Per Selle
Stein Rokkan Centre
University of Bergen
Bergen 5007
NORWAY
Per.Selle@rokkan.uib.no

Øystein Bortne
University of Stavanger
Stavanger 4021
NORWAY
oystein.bortne@uis.no

Kristin Strømsnes
Stein Rokkan Centre
University of Bergen
Bergen 5007
NORWAY
kristin.stromsnes@rokkan.uib.no

Library of Congress Control Number: 2005936369

ISBN-10: 0-387-30524-6   e-ISBN 0-387-30525-4
ISBN-13: 978-0387-30524-0

Printed on acid-free paper.

Printed in the United States of America.   (SPI/IBT)

9 8 7 6 5 4 3 2 1

springer.com

Contents

Chapter 1
In Defense of Uniqueness

In analyzing environmentalism in Western democracies, researchers have paid much attention to similarities across nations and findings common to most countries. This strategy is plausible given the relatively new status of the field of environmentalism. Now that environmentalism is more established, we venture a different approach. We intend to examine the way organized environmentalism, environmental beliefs, and behaviors are configured in a country that does not easily fit into the general assumptions behind the definitions of environmentalism that has been generated by the bulk of "continental" or "Anglo-Saxon" literature.[1] Our intention is to explain and demonstrate that the Norwegian case of environmentalism is unique and anomalous. Our study is also one of a specific political culture with a rather unique combination of state structure and civil society. Organized environmentalism is an important and integrated part of this political culture. More generally, our study addresses the role of civil society and voluntary organizations in a specific type of democracy. In comparing our perspective and findings to studies from other political systems, we want to throw light on why the workings of the Norwegian political system—and, by extension, the Scandinavian political systems—are often misunderstood in the international literature. Without understanding core characteristics of this particular political system and how these characteristics are interrelated one cannot understand the structure and content of Norwegian environmentalism and of Norwegian politics in general.

To appreciate the uniqueness of our Norwegian case and to underscore what we believe is the value of our approach, throughout the book we will relate our perspective and findings to two new and important studies of European and

---

[1] Although continental and Anglo-Saxon countries might differ between themselves as to environmental ideas and perception of the state, the Norwegian case as well as the other Scandinavian cases differ from both sets of countries. The uniqueness makes Norway most different within a broad international context but less different when compared to other Scandinavian countries, whose political institutions are shaped through a braided history. Norway was in union with Denmark and Sweden from 1397 to 1523, with Denmark alone until 1814, and then with Sweden until 1905.

American environmentalism. One study is by John Dryzek and colleagues (2003). This study—*Green States and Social Movements*—is a comparison of the United States, the United Kingdom, Germany, and Norway. The other study is by a team led by Christopher Rootes (2003). In the volume, the authors study and compare seven major countries: Britain, France, Germany, Greece, Italy, Spain (including a separate chapter on The Basque country), and Sweden. One of the fascinating aspects of comparing our study to these volumes, we believe, is that the Dryzek study includes the case of Norway, whereas the Rootes study includes the case of Sweden. This comparison provides us with the opportunity to show that the Dryzek study has gotten the Norwegian case wrong and that Norway has less in common with neighboring Sweden than the Rootes study might lead us to expect. Together, the broad and general perspectives of these two studies give us the opportunity to show more clearly the fundamental features of Norwegian environmentalism. This analysis is necessary and important in order to understand environmentalism more fully.

Specifically, we argue that environmentalism operates differently in Norway compared to other polities. The Norwegian case deviates from the mainstream definition of environmentalism due to two anomalies. The roots of the two anomalies are found at different but interrelated levels of analysis. The first anomaly is that the Norwegian political and social system differs when compared to systems in most other countries in that adversary actors or interests are not excluded from national politics, but, in general, are welcomed by the government. We label this anomaly the *inclusive polity and the state-friendly society*. In short, we refer to it as *the state-friendly, or government-friendly, society*. The anomaly of state-friendliness, which has a long historical tradition, is primarily found at a *structural* level although it also ties in with mainstream attitudes and beliefs that define part of the Norwegian political culture. The other anomaly is more *ideological*, although this, too, has strong structural ties because national environmental concern is influenced by a notion of social hardship and self-sufficiency as part of local living. The essence of this anomaly is the protection of man in nature. This anomaly has maintained its strength because of a relatively low level of urbanization in a geographically elongated and drawn out country. However, this anomaly has a long history too because of the historically strong tradition of local democracy in the Norwegian polity. We label this anomaly the *local community perspective*.

In isolation, the two anomalies are not unique. First, a state-friendly society is characterized by a specific type of close relationship between state and civil society. This societal type is found in the Scandinavian countries only and contributes to explain these countries' specific type of a universally based welfare state regime. This anomaly therefore excludes the vast majority of countries around the world.[2] Second, although the local community perspective might be found

---

[2] Something similar is supposedly found in Belgium and the Netherlands. See also van der Heijden (1997).

elsewhere, it is not found in the other Scandinavian countries to the same extent.[3] We believe that it is the *combination* of the state-friendly society and the local community perspective that makes the Norwegian case unique in the international context. Therefore, we will also discuss how the anomalies interact with each other and are dependent on each other.

We believe that it is important to address deviant cases because most countries to which general environmental theory more or less easily applies ostensibly have less need to account for anomalies. Failing to address unique cases leads to them not being properly understood as well as permitting too much noise in standard explanatory models. In consequence, failing to address deviant cases prevents researchers from pinning down the limits and validity of general theories.

Data

Our analysis of organized environmentalism in Norway takes advantage of a rich cluster of sources. The main bulk consists of comprehensive surveys, personal interviews with key persons in the field, as well as semistructured interviews with the leadership of environmental organizations. The Survey of Environmentalism ("Miljøvernundersøkelsen") consists of two sets of extensive postal surveys carried out in 1995. One survey was administered to a sample of the Norwegian population ($N = 2000$) age 15 and above, randomly drawn by the Norwegian Government Computer Centre from the Central National Register. The response rate of the general population was 52.4% ($N_{GP} = 1023$). The other survey was administered to members of 12 Norwegian environmental organizations, the selection of which had taken place during initial research and several visits to the organizations. These organizations are representative of the environmental field in Norway. At the time of the surveys, their population sizes ranged between 140 members and 180,000 members. The 12 organizations are The Norwegian Mountain Touring Association (Den Norske Turistforening); The Norwegian Society for the Conservation of Nature (Norges Naturvernforbund); Nature and Youth (Natur og Ungdom); World Wide Fund for Nature (WWF— Verdens Naturfond); Norwegian Organization for Ecological Agriculture

---

[3] It is possible that the Norwegian view on nature resembles the view found in Iceland and in the Faeroe Islands. It is also possible that the Norwegian view is comparable with Asian and non-European perspectives. For instance, in forested parts of India (and Tibet), there are wide (nomadic and settled) community interactions with wild animal species. However, the acceptance, at least by the peasantry and rural people, of animals needing "sanctuary space" from which humans should be entirely removed, is, with the objections by urban elites, generally low. We are grateful to Sanjeev Prakash for making the latter point clear to us.

(Norsk Økologisk Landbrukslag)[4]; The Future in Our Hands (Framtiden i våre hender); The Bellona Foundation (Miljøstiftelsen Bellona); Greenpeace Norway (Greenpeace Norge); NOAH—for animals rights (NOAH—for dyrs rettigheter); Women–Environment–Development (Kvinner og Miljø); The Environmental Home Guard (Miljøheimevernet, now called Green Living Norway (Grønn Hverdag)); and Green Warriors of Norway (Norges Miljøvernforbund). (See Appendix B for the organizations' websites.) Approximately 300 respondents were randomly chosen from each organization's member list. From the organization Women–Environment–Development, all 140 members were used. The questionnaires for both surveys were mailed in early May 1995 and the sampling process was closed late June 1995. Funding and national legislation allowed one reminder (by postcard) and two follow-ups to nonrespondents (cover letters with replacement questionnaires). The response rate was 59.2% (N_{OE} = 2088) (Strømsnes, Grendstad, & Selle, 1996). See Appendix A.[5]

In addition to the surveys, the authors organized a conference in November 1995 to which representatives of all 12 organizations were invited. Key personnel of the organizations candidly briefed one another, as well as the authors, on issues like membership policy, organizational model, strategy, and ideology. These proceedings (Strømsnes & Selle, 1996) offer valuable insights to the organizational field in addition to what could be obtained in the membership survey. In 1997, more formal interviews were conducted with the leaders of the organizations. The interviews lasted for about 90 minutes, focusing on a range of topics such as information about employees, income, expenses, membership development, work methods, organizational structure, and organizational decision-making processes. In 2000, leaders of the organizations were interviewed again (Strømsnes, 2001). The surveys, the conference, and the repeated interviews provide a rich empirical base that covers more than a 10-year period of organizational change and stability. Furthermore, this 10-year period is just a smaller part of a longer period during which some of the authors have studied the development and change of the Norwegian voluntary sector at large and the Norwegian welfare state more broadly. This background gives us an opportunity to put the environmental movement and the environmental field into a wider context. Throughout the book, we will therefore compare the development of the environmental movement with the voluntary sector at large. We will also keep an eye for the more or less subtle differences as to how the environmental organizations relate to governmental bodies as compared to voluntary organizations within other policy fields.

---

[4] In 2001 this organization merged with two other organizations to form Oikos–Økologisk landslag.

[5] An English version of the member questionnaire can be obtained by contacting the authors. The population questionnaire contains a subset of the former because many items related to the organizations were omitted.

The Structure of the Book

The book is divided into three parts. In the first part—Perspectives on Unique Environmentalism—in Chapter 2, we elaborate on the anomalies that make the Norwegian case unique. In Chapter 3, we present the 12 organizations that represent the field of organized environmentalism. We present the history of these organizations, the conditions under which they where founded, and how they relate to the state. Then we introduce the two dimensions of organizational age and environmental coreness in order to build a typology by which organized environmentalism can be better understood.

In Part II, we analyze the environmentalists within the context of the two anomalies. We ask who the Norwegian organized environmentalists are (Chapter 4), what beliefs and opinions they hold (Chapter 5), and what their pattern of environmental behavior and political activity are (Chapter 6). In these analyses, we employ a handful of theories, or perspectives, in order to aid our understanding of organized environmentalism. The purpose is not to test these theories as such, but to use the theories as tools in order to grasp meaningful variations of Norwegian environmentalism. Throughout these three chapters, organized environmentalists are contrasted across organizations and organizational types and compared with the general population.

In Part III, we link our survey of the organized environmentalists and the general population with the roots of the two anomalies. The local community perspective makes the conspicuous lack of animal rights in the definition and practice of Norwegian environmentalism more understandable (Chapter 7). Also, the state-friendly-society perspective makes the conspicuous absence of a green party in a green polity less bewildering (Chapter 8). These anomalies make differences in attitudes and behavior between the population and organized environmentalists smaller and less pronounced than what would have been the case were these anomalies not in operation. When environmental organizations operate in a society in which the state and local communities are inextricably tied, organizations themselves become less of an alternative movement and they more quickly develop pragmatic policies. Thus we expect the differences between organized environmentalists and the general population to be rather small and, in some areas, even bordering on insignificance.

In the final chapter, we argue that the context facing voluntary organizations at the turn of a century is quite different from the context of the preceding decades of the 20th century during which most of the 12 environmental organizations under study were founded. At the beginning of the 21st century, four distinct changes apply to the organizations under study. First, old organizations are losing members while more recently founded organizations show a disinterest in recruiting members at all (Tranvik & Selle, 2003; Wollebæk & Selle, 2002a). Second, the increased legitimacy of the market and its actors have led recently founded organizations not only to accept financial contributions from said actors but also to collaborate with them when common goals were identified. Third, comprehensive changes within the public sector itself—often sailing under the name of New Public Management—affect the relationship between the public sector and the voluntary sector and between local and central governments. These changes will put a new type of pressure on these historically rooted anomalies. Fourth, the new information and communication technology, offered through personal computers, cell phone technology, and the Internet, have altered modes of communication within organizations, between organizations and their members, and between organizations and other actors in society.

To speak of a changing society at the turn of a century is therefore less hollow because the rate of societal and technological change is probably faster and more profound than at any period earlier in history and because the change kicks in during the late adolescence of the modern environmental movement. This change will not occur without affecting the nature and operation of voluntary organizations in general and environmental organizations specifically. Hence, our writing of this book is to elaborate on and to understand the nature and relationship between the historically based anomalies of the state-friendly society and the local community perspective as applied to organized environmentalism in this time of transformation. What kind of pressure on the content and structure of the anomalies do we see now? Are they strong enough to transform the Norwegian political culture and thereby Norwegian environmentalism, making Norway gradually more similar to what is found elsewhere? We will relate to these changes throughout the book, because in the longer run they might transform Norwegian environmentalism. In the final chapter, the implications of these changes for the continuing importance of the anomalies in shaping Norwegian political culture are discussed along with a view to challenges to voluntary organizations and their role in a democratic society.

Part I:
Perspectives on Unique Environmentalism

Chapter 2
The Two Anomalies

The Misconceptions of Norwegian Environmentalism

The Norwegian case of organized environmentalism consists of two interrelated anomalies. One is the state-friendly society in which the population holds a basic trust toward state institutions and in which voluntary organizations work closely with governmental bodies. This trust has led to the environmental movement, often understood elsewhere as an alternative movement, having become pragmatic and cooperating very closely with governmental bodies. This mutual cooperation between governmental bodies and voluntary organizations differs in structure and extensiveness across societal sectors, but the cooperation has been very close within the environmental field. The other anomaly is the local community perspective in which animal rights—which is a distinct feature of environmentalism in most other countries—fails to enter the Norwegian definition of environmentalism. These interrelated anomalies make the case of Norway unique within international environmentalism. As this book will show, these anomalies have a profound impact on the size, organization, ideology, work methods, and influence of the environmental movement (Parts I and III). Furthermore, the anomalies also have a great impact on the demographic characteristics, beliefs, and behavior of the organized environmentalists (Part II). Without considering these anomalies, it is not possible to understand the form and substance of Norwegian environmentalism.

If we turn to the international social science literature on Norway, as well as on Scandinavia, the understanding of the political culture being different is nowhere to be found. A consequence of this misconception is that studies of Norway and neighboring countries grasp neither the actual role of government nor the structure and role of civil society, including the role of the environmental movement.[6] One of our aims in this study is to throw light on this misconception by relating

---

[6] Here we will not discuss this misconception, but see Kuhnle and Selle (1992b) and Tranvik and Selle (2005).

our study and perspective to the Dryzek and Rootes studies of modern environmentalism that we referred to in Chapter 1 (Dryzek et al., 2003; Rootes, 2003). We will relate more extensively to the Dryzek-study because its perspective is closer to our study than the Rootes study.

The Rootes volume is an important study that concentrates on the occurrences of environmental protest reported in newspapers in the seven European Union (EU) countries (Britain, France, Germany, Greece, Italy, Spain, and Sweden). The study offers many interesting findings concerning the increased institutionalization of environmental protest in EU countries as well as variations on environmentalism across countries. However, the study holds that the repertoire of environmental collective action is mainly found within the structure of environmentalism itself rather than in the political culture and structure of a single political system. The repertoire of protest is more a matter of movement cultures than national culture. To a certain extent, this might, of course, be true. For instance, independent of country, protest groups against nuclear energy and animal rights are more militant than other types of environmental organization. On the other hand, there are many results that point in the direction of the importance of a country-specific political culture. For instance, the study finds the strong localism of environmental protest in Spain and Greece as part of the strong overall localism in these countries. Also, the study observes the extensive use of environmental demonstration in France, where demonstrations more generally are an important part of the repertoire of collective action. However, in general, the study argues strongly against the idea that environmental protest fits well with national stereotypes, particularly underscoring the extent of violent environmental protest in Britain and Sweden (see especially Chapter 10 of the Rootes volume).

What explains protest then is the combination of a specific movement culture that connects to political conjunctures, or opportunity structures, in the different countries. This is not the least so because protest is connected mainly to the implementation of policies (output response) rather than its formulation (input response). However, this study primarily looks at protest behavior and does not analyze how the environmental movement is structured or how it works more generally in the different countries. For instance, we find almost nothing about the extent of cooperation and how the environmental movement cooperates with governmental bodies or businesses in the different countries. Even if core aspects of environmentalism are not studied at all, one draws very general conclusions about environmentalism and how it should be studied.

In underlining that environmentalism should be mainly understood within the culture of environmentalism itself, the Rootes study also argues explicitly against the Dryzek perspective, underscoring that structural factors, political institutional arrangements foremost among them, explain little if anything of the variation (Rootes, 2003, p. 253). Without addressing or discussing all aspects of the Rootes study's view on politics and society, we do not fully share their understanding of society. Structural factors, not to speak of political institutional arrangements, always count.

Even so, at a more general level and as a next step, the study points in the direction of a need for national studies or case studies. One does not find any strong coordination of environmental activity across countries or a strong transnational influence on strategies and action in the different countries. Furthermore, the EU as an institution still plays a very subordinate role in giving direction to environmental protest. Protest mainly reflects conjuncture of national politics, but it is, in general, becoming more formalized and centralized than before, in which each and every protest action receives less support and less participation. As we will see, particularly in Part II of this volume, to some extent this also fits the Norwegian situation.

The Dryzek study, on the other hand, sees environmentalism through the lenses of national political structure and national political culture. This study tries to identify which historical features cause environmentalism to work the way it does within a particular polity and then compare results with what is observed in other countries. Even so, we argue that Dryzek's study of Norway offers little insight into the structure of Norwegian environmentalism. Specifically, we hold that the Dryzek study supports a rather common misunderstanding of Norwegian and Scandinavian politics.

The perspective of the latest Dryzek study builds upon and expands an earlier study by Dryzek (1996) in which Norway is also included. In the 1996 study, Dryzek argues that Norway is a corporatist system in which most groups, except business and labor unions, are excluded from state councils and, hence, from political power (Norway scores high on studies of European corporatism; see Chapter 9 of this volume). In the 2003 study, the Norwegian political system becomes somewhat more open and inclusive. A prime example is how the environmental organizations cooperates with, or are co-opted by, the state. Here, Norway is now defined through the concept of *expansive corporatism*. We agree in viewing the Norwegian system as open and inclusive. However, we believe that the Norwegian polity still is much more open than what the Dryzek study concludes. Furthermore, we strongly disagree with Dryzek on what this openness actually means. This difference in understanding has, as we will soon see, profound consequences for the understanding of the relationship between governmental bodies and environmental organizations. Furthermore, only looking at how the environmental organizations interact with governmental bodies and excluding the local community perspective as a defining part of Norwegian political culture give, at the end of the day, a somewhat skewed picture of Norwegian environmentalism.

Let us make clear why the Dryzek study is not helpful in our attempt to analyze Norwegian environmentalism and the state–civil society relationship more generally. In trying to grasp the structure of environmentalism in the four selected countries, the Dryzek study starts from two main dimensions: whether the state or government bodies are exclusive (i.e., opens up for a few organizations only) or inclusive (i.e., opens up for many organizations), and whether the state is active or passive when it comes to connecting new organizations to the state. Norway represents the combination of an inclusive and active state, the United States is a passively inclusive state, Germany combines passive and exclusive states, whereas the United Kingdom is the actively exclusive one (even if having periods best characterized by actively inclusive).[7] Whereas the United States was an environmental pioneer around 1970, Norway gradually took over and became the greenest of states. However, one might argue that Norway has not become any greener since and is not expected to become so either because the absence of strong subpolitical groups will not put pressure on this specific type of state. It is a system that in many ways has peaked when it comes to environmental modernization. Dryzek and colleagues argue that a transformation to a green state is now most likely to take place in Germany.

At a general level, we agree on characterizing Norway as an inclusive and active state. The Norwegian environmental movement cooperates closely with governmental bodies, not the least with the Ministry of the Environment (see Chapter 3 of this volume). Several important and, in our view, correct observations are to be found in the Dryzek study; for instance, the observations that subpolitical groups are weak or absent and that the Norwegian environmental movement in a comparative perspective is small both in numbers and activism. This is even more true if one takes into account the extensive voluntary sector found in the country (Sivesind, Lorentzen, Selle, & Wollebæk, 2002). However, this organizational weakness is only true if one looks at the main or typical environmental organizations only. In Norway, one generally finds a very broad definition and understanding of environmentalism, including outdoor recreation and the preservation of cultural heritage (see Strømsnes, 2001). This is also made clear when we study what areas governmental environmental bodies are meant to cover (Bortne, Selle, & Strømsnes, 2002). Furthermore, environmental concerns also play an important role in many voluntary organizations that are not typically being considered as environmental organizations. All in all, even if environmental concern is moderate in character, it is much broader and goes deeper than can be read from the Dryzek study. The extension of environmental concern has an influence on the political space that is available to the more specific environmental organizations.

However, where we disagree with the Dryzek study is in its understanding of the relationship between state and civil society in Norway. The Dryzek study is mainly theoretically based and its general understanding of what a "good" relationship between state and civil society should look like does not work well in a Scandinavian context. Heavily influenced by a corporatist understanding of

---

[7] Even if Sweden is not part of the Dryzek study, in the more general discussions in the book Sweden is placed in the same group as Germany, as a passive and exclusive state. When criticizing the Dryzek study, the Rootes study places Sweden as an actively inclusive state. Even if this is not the place to go deeply into structural differences between Sweden and Norway, we believe that here the Rootes study is closer to the truth. When it is so easy to place one and the same country in different "boxes," the criteria for placing them are not fully specified.

Norway and a Habermas-inspired understanding of civil society, the Dryzek study seem to argue as follows: In Norway, the state is so strong that your organization depends on the state for legitimacy and finances. In this process, your organization becomes co-opted by the state and loses its autonomy. The state cultivates groups that moderate their demands in exchange for state funding and guaranteed participation in policy making. Environmental groups become arms of the state. Furthermore, because of government's extensive use of committees behind closed doors, the impact of ordinary members is weak or nonexistent. What the Dryzek study calls the case of "weak ecological modernization" is understood mainly as a top-down project without grassroots influence.

Here is not the place to go into a comprehensive discussion of the character of the modern state. However, in the Dryzek study, there is a very static and general understanding of the state as monolithic and always the dominant actor. Several studies have shown the Norwegian (and Scandinavian) state(s) as segmented and/or even more fragmented than ever (e.g., Østerud et al., 2003). What these studies show is that different governmental bodies think differently and do things differently, often accompanied with little or no communication between them. This is an indication of weak horizontal integration. The Dryzek study's monolithic understanding of the state, we argue, means that you are unable to grasp the dynamics in the relationship between civil society organizations and the state both generally and within different policy fields. The organizations are not always dominated by the state, but in some policy fields, domination happens more than in other fields. In general, in Norway, the organizational input and autonomy are stronger than the seemingly closed theoretical system and bird's-eye view of Norwegian politics that the Dryzek study uses and takes into account.

Additionally, the lack of a more general understanding of Norwegian political culture makes things worse because the local community perspective is not taken into account. Their cooptation argument starts from the view that environmentalism has a deep structure or a core that is quite radical, but that it becomes moderated by being co-opted by unfriendly state institutions. For the case of Norway, we argue the opposite. Because of the overall political culture that environmental organizations are part of and operate within, we argue that environmentalism in general is moderate from the start even if the movement has proved to be more radical in some periods than in others (see Chapter 3 of this volume). That goes both for the individual environmentalists as well as for the environmental organizations. Within this system, furthermore, the organizations are more autonomous than what the Dryzek study allows. For instance, studies have shown that governmental bodies never interfere with internal organizational processes (Selle, 1998). Such interference would be to overstep a line that has been in existence for a long time. Altogether, these are not small points because it means that we see the Norwegian political culture and not the least the state–civil society relationship as fundamentally different from that of the Dryzek study. A further implication of this difference is that we understand Norwegian environmental policy less of a top-down government project (even if that is the case in many other policy fields). These differences in perspective seem to be so profound that we, in effect, perhaps are talking about two fundamentally different types of democracy.

Because of what is seen as a lack of organizational autonomy and political subgroups with members or grassroots having little influence in organizations and society, the Dryzek study seems to conclude that Norway lacks an active oppositional public sphere or a vibrant civil society. This Habermas-inspired understanding of the public space in combination with a monolithic understanding of the state, in which the state is always the dominant part, makes the Dryzek study conclude that Norway is a "thin" democracy. In what follows, we argue strongly against this understanding of too much state and too little civil society. Historically speaking, Norway is a "thick" democracy with a vibrant civil society. In a comparative perspective, this vibrancy becomes even more pronounced (Salamon & Sokolowski, 2004; Sivesind et al., 2002).

However, as we will return to later, particularly in Chapter 9, the transformation of the voluntary sector and of the state structure that takes place now might change the relationship between the state and civil society in fundamental ways. Perhaps one consequence of these changes could be that the perspective of the Dryzek study, even up to now being so off the mark, in the future can offer insights into the Norwegian and Scandinavian politics. However, we are far from reaching that point yet. In the meantime, let us take a closer look at the characteristics of the two anomalies that make Norwegian environmentalism unique.

An Inclusive Polity in a State-Friendly Society

The strongest political support for the active and interventionist Norwegian state has historically come from the politically broad and popular center-left. Since the late 19th century, the Norwegian state has built on popular movements and mass parties, including the strong political position of the labor movement after the Second World War. Public ownership of land, resources, and capital has been extensive, whereas the private economy has been geographically decentralized with relatively small enterprises. Through the universalized welfare arrangements and other public institutions, like the official Lutheran church with a status as a comprehensive popular church, the state has gained wide support and legitimacy. Mass parties and voluntary associations with a broad societal agenda have made a strong impact on the development of public authority. The Norwegian nation-state, as a political community, has been a framework for popular participation and social and regional redistribution.

This means that Norwegian—and Scandinavian—politics in a comparative perspective might be characterized by high levels of institutional centralization and state friendliness (Kuhnle & Selle, 1992b). It is, for instance, the state—

rather than markets, religious institutions, or local community associations—that has been the paramount agent of social and economic reform, most notably the development of comprehensive welfare schemes and the system of corporative economic planning, with the state as the most important participant. It is therefore tempting, especially, perhaps, for political scientists of Anglo-American origin, to conclude that in the Scandinavian countries, there is too much trust in and too much dependence on the state bureaucracy while too few checks and balances limit the scope of state power. Even if it is hard to brush off these (and similar) criticisms of "the Scandinavian exceptionalism" as misguided, they tend, nevertheless, to be informed by simple and so-called protective models of democracy (Held, 1996) by which the Dryzek study is also influenced.

According to this model, institutional centralization and state friendliness are at odds with the notion of democracy because democracy works only when power is decentralized, when citizens are legally protected from being within the reach of the "tentacles of the state," and when everyone is free to carry out his or her own life plans as they see fit. The challenge of democracy, as viewed through the lenses of the protective model, is therefore to encroach the exercise of state power. In Norway and Scandinavia, however, the democratic challenge has been perceived rather differently. Institutional centralization is not regarded as a problem, so long as there are ways for ordinary citizens to influence the exercise of state power. State-friendliness is seen as the clearest manifestation of the democratization of centralized state power. Citizens view the state benignly because through ideological mass movements they have been thoroughly plugged into the running of the state (Tranvik & Selle, 2005; Wollebæk & Selle, 2002a). It is this social contract—high levels of institutional centralization balanced by high levels of citizen control—that is now being gradually eroded (see Chapter 9).

The history of West European states is a history of conflicts between, and integration of, different groups and classes in society. Conflict integration might be seen as consisting of four thresholds, like locks in a canal: legitimation, incorporation, representation, and executive power (Rokkan, 1970). The passing of each will gradually take a political movement closer to the pinnacle of the polity. The conflict-integrating mechanism is not a static dualism where a movement is either in or out. Rather, it can be seen as a continuous process where the movement struggles for increasing power and influence. In a comparative perspective, this process has been rather smooth in Norway. In the course of history, the power and position of the central establishment were challenged by groups that wanted to put their mark on the development of the state and the nation. Different movements, such as farmers, radical intellectuals in the cities, countercultural lay-Christian, teetotalists, linguistic movements in the periphery, and, finally, workers, were organized and mobilized to push forth the interests of the group. At the end of the day, members of a social movement might have been given the right to vote and representatives of the political wing of a social movement might have gained representation in parliament or, lo and behold, even obtained access to the executive power.

The prominence of social movements, which we believe is not fully outlined by the Dryzek study, can be considered a common feature of the Scandinavian countries.[8] In the crucial junctions that formed the history of the modern Western European states, the roads traveled by the Scandinavian countries are astoundingly alike (see Rokkan, 1970, especially Chapter 3). Consequently, the structuring of mass politics in Norway, and hence modern Norwegian politics, is a result of mass movements' and voluntary organizations' struggle for representation and power. The people's movements have had a prominent, even mythical, place in the minds of Norwegians. Voluntary organizations have been important for the development of democracy and for the nation-building process. In the Scandinavian countries, the organizations recruited members from a *broader* social basis than in most other countries (Sivesind et al., 2002; Wollebæk & Selle, 2002b). Through the incorporation of broad member-based organizations, the national and the local political levels were linked closely. The close ties between state and local government reinforced this closeness (Tranvik & Selle, 2005).[9]

Hence, the links between state and voluntary organizations have been many and dense since the growth of a separate voluntary sector from the mid-1800s (Kuhnle & Selle, 1992b; Selle, 1999). Voluntary organizations turn to the state for cooperation, funding, and legitimacy. As a consequence, an organization that seeks influence in the political process must turn *to* the state and not *away* from it. Cooperation with the state means that the organization increases its political influence, that it increases its legitimacy among the public, and that the organization is far more likely to receive financial contributions from the state (Selle & Strømsnes, 1996; Tranvik & Selle, 2005). However, it does not follow that the organizations should be discouraged from criticizing the state. To the extent to which these organizations are watchdogs of the state, it is odd that the "watchdogs are also fed by the one they are intended to watch"

---

[8] Like so much of the welfare state literature, the Dryzek study also put too much emphasis on the mobilization of the working class, enabling the entry of this organized class into the state, as the defining force behind the growth of the welfare state. However, neither the structure nor the content of the different welfare states can be understood properly if researchers are not sufficiently perceptive of the impact from other social movements too. Not the least in Norway has the broad impact from different social movements in this inclusive state been of great importance. For a broad discussion of this, see Berven and Selle (2001) and Kuhnle and Selle (1992b).

[9] We have chosen to use the concept "voluntary organization" and "voluntary sector" in favor of concepts like the "third sector," the "independent sector," and the "nongovernmental sector." The reason for this is that the term "voluntary" "[…] tells us something about both the members in the organizations, members are not forced into membership, and about how the organizations came into being. […] they are not forced into existence." (Kuhnle & Selle, 1992a, pp. 6–7). In addition, the concept "voluntary organization" has been used in colloquial speech in Norway in most of the 20th century (Sivesind et al., 2002).

(Tjernshaugen, 1999, p. 39, our translation). In this process, contrary to what one observes in other countries, organizations have *not* emphasized the importance of defining or pursuing a sphere autonomous or independent of the state.[10] Such a sphere has been more important in most other countries, including the three countries other than Norway in the Dryzek study.

However, even if the environmental movement is heavily dependent on state for financial support if they want to keep the level of activity they have become used to, one should not overemphasize the financial dependence. The Norwegian voluntary sector is less dependent on governmental financial support than in most West European countries. In general, it is the service-producing organizations within the health and welfare field that are the most dependent, whereas cultural, leisure, and advocacy organizations get a smaller part of their overall budget from the government. In general, only 35% of the revenues of the voluntary sector in Norway comes from the public sector, compared to 55% in the EU countries (Sivesind et al., 2002). This important structural feature does not seem to fit well with the assumptions underlying the Dryzek study. The close interaction with government within the environmental field is probably easier to understand if we take into account the scientific heritage of the environmental movement in which the integration into governmental bodies was high and membership was extremely low until the end of the 1960s (see Chapter 3).

The development of relations between voluntary organizations and the Norwegian state is often one of rather tight integration. To better understand the structure of this relationship, let us take a brief look at the health and welfare sector in which the voluntary organizations have been particularly strong. In this sector, voluntary organizations cooperated closely with the state in the period between the two world wars. These organizations were almost *sine qua non* for turning health into a public issue (Berven & Selle, 2001; Kuhnle & Selle, 1992a, 1992b). The voluntary health organizations' strategy was to cooperate constructively with the public authorities and to press them to take public responsibility. This development was characterized by harmony (rather than conflict), consensus (rather than ideological disagreement), cooperation and division of labor (rather than isolation and segmentation), and, hence, by mutual dependence (rather than one-sidedness in the economy, in work resources, and in legitimization) (Klausen & Selle, 1995). Because of the early organization of health as a public issue, its concurrence with economic growth, and its importance to the population at large, health and welfare have become the most integrated issue in the highly developed welfare state. As environmental concerns became politically important from the late 1970s and early 1980s, something quite similar happened to this policy field. However, the role of the voluntary organizations was even more important for the development of the welfare state than for the development of the environmental

---

[10] However, from the late 1990s we do see a gradual development in this direction (Sivesind et al., 2002).

field.[11] However, the structure of cooperation is part of a long tradition. Historically speaking, in this country it is the way of doing things.

The integrated participation between voluntary organizations and the state has two implications. Because of public financing and government backing, state proximity might be necessary for organizational survival. The advantage for the organizations is increased influence over policy, efficiency, and legitimacy. State proximity is not the problem but the *solution* for an organization whose interest is more than sheer survival. However, because the voluntary organizations are being tied to the state, there is also a price to pay, such as responsibility, some loss of autonomy, less ideological purity, and that the organization also has to take into account the demands from the state, not only those from its members. Taken together, this might create a dilemma for the organization, because evidently "benefits seem to be inextricably bound together with costs" (Olsen, 1983, pp. 157–158). However, as we will see, not only the environmental organizations but also the rest of the voluntary sector and the population at large see the benefits as greater than the costs. It is not something into which the organizations are forced.

The intimate relationship between state and organizations violates a liberal understanding of the tripartite, power-balanced relationship among state, market, and civil society. Within this perspective, voluntary organizations have to be autonomous whereby they constitute the core of a free and independent civil society (i.e., the protective model of which the perspective of the Dryzek study is a part) (Held, 1996).[12] The German philosopher Jürgen Habermas (1987) argued that society in general can be divided into two large spheres. The system world sphere consists of the political system (the state) and the economic system (the market). The life world sphere is the civil society. This thought is further developed in Cohen and Arato (1992), in which the civil society is a sphere for social interaction between the state and the market. According to them (and many others), it is especially important that actors in the civil society can influence the political sphere without being integrated into political and administrative bodies, thus making the civil society, *and* the organizations, autonomous (Bratland, 1995, p. 19; Tranvik & Selle, 2003). It is evident that the Norwegian case fails to meet Cohen and Arato's criteria. However, that does not at all mean that, in a comparative perspective, we are talking about a "thin democracy" as the Dryzek study does. As comparative studies show, Norway has one of the most extensive and dynamic civil society there is (Salamon & Sokolowski, 2004; Sivesind et al.,

---

[11] This might in part be explained not only by the tremendous strength of the voluntary sector in this field (Kuhnle & Selle, 1992a) but, following the argument of the Dryzek study, also by the fact that the welfare state is a state imperative while environmentalism so far is not. That is why the Dryzek study argues that there is still no green state, even if expecting the modern state to develop in that direction.

[12] This theoretical tradition dates back to Tocqueville, Locke, and Mill. See also Nisbet (1962) and Berger and Neuhaus (1977). For a critique of this theory, see Salamon (1987) and Tranvik and Selle (2003).

2002). The historically important role of civil society in many ways constitutes Norwegian democracy, which is a system that is especially open to civil society input from democratically built organizations with strong local branches. The Dryzek study, we believe, fails to include this important feature.

The intricate ties among state, market, and organizations affect political decision-making. Changes in the Norwegian society, and in the rest of Scandinavia, are mostly the result of what is referred to as "considered reforms," in which a large number of different organizations are involved in the not uncommonly protracted and open hearing processes.[13] However, in their study of the reform processes in Sweden, Brunsson and Olsen question the freedom of choice of the reformists, stating that "reforms are difficult to decide upon, to execute, to get the desired effect of, and to learn from" (1990, p. 13). One can spot the failure of many reforms in the necessity of reforming the reforms (Brunsson & Olsen, 1990, p. 255). One possible reason for the alleged limited power of the reformists is that when all of the powerful interest groups are participating in the decision-making process, the result is perhaps not what is best for the reform itself, but, rather, what is best for a compromise among the actors in this pragmatic political culture. The broad understanding of environmentalism within the Norwegian political culture gives room also for organizations that are not strictly environmental. This has a further moderating effect on environmental policies.

If we place voluntarism in a comparative perspective, Norwegian voluntarism is characterized by a high degree of membership but with an extensive share of passive members (Dekker & van den Broek, 1998, p. 28). Norway, Sweden, and The Netherlands have a very high level of voluntary affiliations and memberships (Sivesind et al., 2002; Wollebæk, Selle, & Lorentzen, 2000, p. 27). Almost three-quarters of the general Norwegian population between 16 and 85 years of age is member of at least one voluntary organization. Each person is, on average, a member in approximately two voluntary organizations. Among members only, the average number of memberships is 2.4 (Wollebæk et al., 2000, p. 52).[14] It is within this context of an extensive voluntary sector that it becomes interesting that the environmental organizations have such low membership figures (see Chapter 3). In general, due to the prominence of voluntary organizations, citizens often consider it important to be a member. As for the organizations themselves,

---

[13] An example of openness and state–organization linkage is found when the Ministry of Agriculture permits NOAH—for animal rights, one of the most radical environmental organizations in Norway, the opportunity to inject the organization's view into the state's policy on animals (Martinsen, 1996). Another example is the radical Norwegian women's shelter movement, which managed to make its cause a public issue. The organization has been able to cooperate closely with state authorities without compromising its leftist ideology (Morken & Selle, 1994).

[14] This number is probably too low, as respondents tend to forget memberships in broad associations such as the Norwegian Automobile Association (800,000 members), the Norwegian Air Ambulance Association (410,000 members), and housing cooperatives (Wollebæk et al., 2000, p. 41).

many members make an organization more legitimate because it can claim to represent large groups in the society. High membership figures make it easier for the environmental organizations to cooperate with the state, at least up to the mid-1980s, when a new generation of organizations emerged. In addition, membership figures are often used to determine the amount of state financial support (Bortne et al., 2002). However, because of low membership figures, environmental organizations become dependent on governmental financial support and, particularly, project support, of which governmental bodies have a direct interest in the outcome. A consequence might be that the organizations look more for governmental project money than for new members in securing the organization financially. This transformation of modern politics might, to a certain extent, explain the low membership figures.

Much of the citizens' trust and the organizations' trust in the state are accumulated in the process by which next to any group can be consulted in the state's decision-making processes. The continuous conjunction between an inclusive polity and a state-friendly society yields a special structure by which the polity and the society grow even closer in an intricate net. The relationship between the two has, up to now, been based on mutual confidence and trust. This, however, does not at all exclude the possibility of disagreement. We are not in a heaven of harmony and, as we show in Chapter 3, periods of rather deep conflict have occurred. However, with the introduction of New Public Management ideas and tools in the public sector, we might now see a change from a trusting relationship to increased governmental control. These ideas and tools emphasize cost-effectiveness and "contracting" at the cost of trust. This is a process that might gradually transform the political culture itself.

Evidence of Norway being a state-friendly society in Europe can be found in the level and rank of its citizens' trust in institutions and in social capital in general (Wollebæk & Selle, 2002b). The European "Beliefs in Government" study shows that Norwegians' trust in political institutions (i.e., the armed forces, the education system, the legal system, the police, the parliament, and the civil service) were highest both in 1982 and in 1990. Seventy-six percent and 68% of the respondents had either "a great deal" or "quite a lot" of confidence in these institutions in 1982 and 1990, respectively. In Britain, for instance, the figures were 64% and 58%, respectively. Norwegians ranked second only to Ireland on trust in more private institutions (i.e., the church, the press, the trade unions, and major companies). On a generalized trust score, Norway ranked number one at both points in time (Listhaug & Wiberg, 1995). More recent and comparative studies show that trust is still high despite a weak decline in some of the measures (Listhaug, 2005; see also Chapter 8 of this volume).

This discussion suggests that state-friendliness consists of two dimensions. *Dependency* on the state varies according to what extent the state is able to control the organization's finances and whether organizational legitimacy depends on the state. *Proximity* varies according to the scope, frequency, and easiness of communication and contact between the organization and the state (Kuhnle & Selle, 1992a). In the Norwegian case, most voluntary organizations that are both close to and dependent on the state are not necessarily dominated by the state. As we will discuss in the next chapter, the state and environmental organizations have, for a long period of time, moved closer to one another. This does not imply, however, that the state penetrates the organizations and takes control over their internal life, as implied by the Dryzek study. Norwegian voluntary organizations have a long tradition of internal organizational autonomy that is also a defining part of their self-understanding (Selle, 1998). Due to the close relationship between the state and organizations, the state is also influenced by the organizations. With the exception of foreign aid, in no other field is that more true today than within the environmental field, in which we have had rather professionalized and scientifically based organizations working in close cooperation with the Ministry of the Environment and other governmental bodies (Bortne et al., 2002). However, as the environmental policy era becomes increasingly mature and institutionalized, perhaps the role of the environmental organizations becomes more one of implementing public policies at the cost of influencing the decision-making itself (Tranvik & Selle, 2005).

This cohabitation between the state and civil society, deeply embedded in the political culture, is a more sophisticated relationship than one of state domination only. Without the organizations, governmental environmental policies would have been less extensive and structured differently and the role of environmental thinking a less important part of the public discourse. Within another type of state, the organizational form, ideology, and repertoire of collective action of the environmental movement would have looked different too. That is why we so strongly argue for the understanding of the dynamics of the relationship between state actors and civil society organizations. However, in the case of Norwegian environmentalism, this relationship takes place within an important policy field in which the local communities are of particular importance.

The Local Community Perspective

Roughly two-thirds of Norway is mountainous and some 50,000 islands lie off its much-indented fjord-frequent coastline. The country combines a vast wilderness with a sparse population. For centuries, in a land with an often inhospitable climate, the inhabitants made a living where land could be cultivated, game could be hunted, and fish could be caught. Traditionally, rural inhabitants have balanced between fighting against the seasonal wild forces of nature and harvesting from nature. This way of living developed strong ties to nature and nourished the national ideal of the local self-reliant community. In this view, nature must be husbanded and not exploited because life in local communities never easily permitted families to leave one place to move to another unsettled place. A rational harvest of nature is not only acceptable, it is the only viable relationship between humans and nature.

Norway had been a strong and unified nation-state in the Middle Ages. Later, under Danish rule, it suffered—tongue in cheek—a 400-year eclipse. The so-called suffering entailed that Norwegian farmers retained their freedom and independence on family-owned farms and experienced less repressive taxation compared with farmers and peasants in continental Europe. Norway never had farmland sufficient to support a landed aristocracy or strong and wealthy urban elites. Because the cities were considered infected by Danish and aristocratic values, the roots of the nation's independence in the early 1800s were sought in the Norwegian countryside, which had never been suppressed by feudalism. Due to the historical weakness of Norwegian urban elites, the Norwegian periphery was never strongly subjected to the cities.

The free Norwegian farmers not only bridged the independence in 1814 with the strong Norwegian nation-state in the Middle Ages, they also accorded the peripheral rural areas a high degree of legitimacy.[15] In a sense, independent Norwegian farmers fueled a certain antiurban sentiment and tension in the Norwegian society. Cities have been regarded with considerable skepticism. Urban movements have never gained any kind of momentum against the rural periphery, and it is first during the 1990s that we can see the contours of a more specific urban policy. On the contrary, "opposition to central authority became a fundamental theme in Norwegian politics" (Rokkan, 1967, p. 368). Indeed, the center–periphery conflict is constituent for Norwegian politics, along with the left–right and the cultural/religious cleavages (Flora, 1999; Rokkan, 1967, 1970; Tranvik & Selle, 2003; Østerud et al., 2003). Even within a rather centralized unitary state and contrary to what the Dryzek study implies, this means that a strong grassroot-based and politicized civil society has received input from below. This has contributed to a strong cross-level integration.[16] However, when we look closer at the environmental movement, a somewhat paradoxical situation emerges. Compared to the voluntary sector in general, the environmental organizations have been more centralized and more professionalized and maintained a weaker organizational base at the local level. Even so, a strong cognitive or ideological orientation toward the local level still exists (see also Chapters 3 and 5).

Today, Norway has a population of 4.5 million people. It has a density of 13 people per square kilometer, which is among the lowest in Europe.[17] Many small communities are found at the bottom of remote fjords and on remote islands. Comparatively, Norway is still more of a rural country than most European

---

[15] It is interesting to note that in a country almost without farmland, farmers are held in high regard in contrast to fishermen, who have been held in somewhat lower regard despite oceans that never have been in short supply (see also Sørensen, 1998).

[16] With the transformation of the voluntary sector now going on, these features are weakened. The grassroot influence is, in general, weaker and the voluntary sector has become less political in character (Tranvik & Selle, 2003; Wollebæk & Selle, 2002a).

[17] For example, the figure for Sweden is 19, 104 for France, 118 for Denmark, 217 for Germany, and 238 for the United Kingdom (Castello-Cortes, 1994).

countries even if the proportion of people living in so-called urban areas is 75%.[18] However, one should keep in mind that "urban" in a Norwegian setting does not really mean urban living similar to what you find in large parts of the European continent. More than 77% of the 434 Norwegian municipalities have less than 10,000 inhabitants. Only nine municipalities have more than 50,000 inhabitants, their total population of 1.3 million inhabitants being equal to the total population of the 337 smallest municipalities.[19] The largest Norwegian cities are rather small in a European context. The capital, Oslo, has approximately 500,000 inhabitants (with an addition of 1.5 million inhabitants in the wider southeastern region). The second, third, and fourth largest cities range between 222,000 and 103,000 inhabitants.

With a large territory and a dispersed population, small and medium-size cities in Norway often find themselves as asphalt islands in a rural sea. As a consequence, city dwellers often find the travelling distance between city life and untouched nature comfortably short. Nature is found immediately outside the city limits. In addition, a late but incomplete urbanization has resulted in a high degree of city residents being able to recount close ancestors whose lives or outcomes are or have been based on farms. Three out of four urban (i.e., cities, suburbs, and towns) residents report that they, their parents, or their grandparents have lived on a farm (see also Chapter 7).[20] One consequence of the frequency of these rural roots is that, cognitively, nature's primary basis of livelihood is difficult to uproot. In addition, city dwellers often take advantage of recreation in nature. This accounts for the alleged puzzle that city residents still can hold a genuine rural and local orientation.

Man's adaptation to living in rugged nature and the egalitarian and rural roots of national identity provide the foundations for what we call *the local community perspective,* the essence of which is *the protection of humans in nature* (see Kvaløy Setreng, 1996, and Chapter 7 of this volume). The local community perspective, which in combination with the state-friendly society constitutes a central part of Norwegian political culture, has a number of corollaries.

Modern Norwegian environmentalism has always been oriented toward local communities and even more so than many other types of voluntary organization (see Chapter 5). Large parts of the Norwegian environmental movement, especially the organizations that emerged at the end of the 1960s and the beginning of the 1970s, have their roots in "Norwegian populism." The term was coined in 1966 by the social anthropologist Ottar Brox, whose political thinking strongly influenced the regional development program of the Socialist People's

---

[18] Again, the figure for France is 74, 84 for Sweden, 86 for Germany, 87 for Denmark, and 89 for the United Kingdom (Castello-Cortes, 1994).

[19] These numbers have shown a remarkable stability (SSB, 1975, 1985, 1995).

[20] Source: Survey of Environmentalism.

Party (the predecessor of the present Socialist Left Party) (Gundersen, 1996; Sætra, 1973). In the analysis of what was claimed to be a failed development in northern Norway, Brox (1966) advocated that the only way out of this economic impasse was found in a small-scale local orientation. A populist, Brox claimed, must understand the society in the northern region as a merger of local communities, which, in turn, consist of a merger of families. Thus, in order to develop this northern region economically, one must start with economic development in the local communities, which, in turn, implies maximizing the economic possibilities for each family. Hartvig Sætra, a self-declared populist, linked the populist alternative more strongly to environmentalism than did Brox. The hope for the future in the populist alternative is only found in the return to the local community and the local economy (Sætra, 1973).[21] The close link between populism and environmentalism in Norway is also observed by Andrew Jamison, who, in his comparison of environmentalism in the Scandinavian countries, stated that "in the Norwegian case the environmental engagement has been followed by a down to earth populism" (Jamison, 1980, pp. 108–109, our translation).

The cognitive orientation toward what is local has traditionally been relatively strong within most environmental organizations, as it is in the voluntary sector in general.[22] Not only do we find a strong emphasis on what is local also in voluntary organizations mainly working at the national level, but institutionally, we also see another important structural feature that explains the strength and continuity of the local community perspective. There is a strong tradition of local democracy where municipalities have retained autonomy from the state on important matters. Despite this tradition, local autonomy has decreased over the last 20 years.

Notwithstanding being a unitary state, the Norwegian system of government must be characterized as relatively decentralized. The municipal level is providing many of the most important welfare services and local governments have traditionally held the power to adjust national welfare schemes to the local conditions. In addition, the Norwegian Municipality Act of 1837 for a long time held a special position in the collective Norwegian consciousness. In the peripheries, it established and institutionalized local self-rule through democratic elections. This self-rule has so far not been a smokescreen. It was not a smokescreen where local communities simply implemented public policies that had been decided at the top of the political food chain. This local autonomy is not only important for the survival of the local community perspective but also for the survival of the state-friendly society. Government is not something distant. It remains close to the inhabitants because it actually takes care of tasks that are

---

[21] Illustratively, the cover of Sætra's book depicts a farmer with a plough.
[22] Although local orientation is common within voluntary organizations in general (Wollebæk et al., 2000), it is our impression that the local orientation is both stronger and more ideological within the environmental sector (Selle, 1998; Tranvik & Selle, 2003).

important for our daily life (welfare, education, social security, etc.).[23] This political arrangement implies that a possible weakening of the local community perspective and of local government might, in the long run, have consequences also for the amount of state-friendliness.

The local orientation has been prominently present in the organization The Future in Our Hands. Steinar Lem, an information officer of the organization, identified the term "being local" as highly honorable in Norwegian environmentalism (Lem, 1996). However, he also warned against being too local, thus becoming too small and insignificant. To act on their local ideals, Nature and Youth's experiment with "democratic decentralization" led to them abolishing the central level of their organization at the end of the 1960s, only later to admit that the attempt was fruitless (Persen & Ranum, 1997). We find this local orientation within most of the other organizations too (Strømsnes, 2001).

A consequence of the local community perspective is that nature does not become a museum of unused or unspoiled nature. Rather, it is a territory designed for the benefit of human beings. For instance, the defiant Norwegian views on whaling must be understood in both a historical context and a local context (see Chapter 7).[24] The support for small-scale, local-community-based whaling, and seal hunting too, is based on an organic way of life in which the local community is linked to nature through its use of the resources conferred by nature. Thus, the supportive and mainstream Norwegian opinions of whaling and sealing should be understood as protection of Norwegian local communities as well as a rational harvest of nature. One can also view Norwegian whaling as a symbol of independence and self-determination. It is difficult for the central government to bypass the local government in these matters. True, as a country, Norway does not depend on whaling. However, there are still small communities where whaling makes an important contribution to the local economy.

Another consequence of the local community perspective is found at the level of national policy on predators. Norwegian predators include the wolf, bear, lynx, and wolverine.[25] During the summertime, many farmers in the southeastern parts of the country allow their herds of sheep to graze in unfenced parts of nature. Unsurprisingly, here the sheep are easy prey.[26] Because the predators seriously interfere with the livelihood of farmers, it is maintained that the predators should be killed or, alternatively, especially the wolf, be firmly relocated in neighboring

---

[23] For an interesting discussion of the role of central and local government in these important matters, see Strandberg, 2006.

[24] Norwegian whaling was a large industrial business in the Antarctic area from the beginning of the 20th century. Such whaling is significantly different from whaling based in or strongly linked to local communities.

[25] In the mid-1990s, the quantities of these animals were estimated to be 20–40 wolves, 26–55 bears, approximately 600 lynxes, and 130–190 wolverines (Knutsen, Aasetre, & Sagør, 1998, p. 64; Miljøstatus, 1999).

[26] The Sami population in the northern part of Norway has had the same problem concerning their reindeers.

Sweden, where there is even more unpopulated wilderness. Wildlife preserva-
tionists argue, however, that predators should not in any way be removed from
their natural habitat and that it is the responsibility of the farmers to keep their
sheep away from predators, who only follow their natural instinct.

In the winter of 2001, the then minister of the environment, Siri Bjerke, ordered
hunters to track down and kill 15 wolves in Norway. The hunters had a field day
because they were permitted to use helicopters during the hunt.[27] The justification
behind this policy was to protect farmers' livelihood in nature and, furthermore,
to secure local influence on the local decision-making process. There have been
strong conflicts between central and local governmental bodies in these matters.
Whereas the central level seeks to balance the interests of wildlife and local communities, the local government argues for having the right to decide itself, almost
always deciding to the benefit of the farmers. Environmental organizations,
except for the World Wide Fund for Nature (WWF), have been very passive in
these matters. This passivity would have been difficult to understand had it not
been for the local community perspective. We cannot always decide whether this
passivity stems from not really being interested in the predators or whether it is
better explained by being afraid of coming into conflict with local interests. For
example, Nature and Youth has had considerable cooperation with the
Smallholder Union. All in all, there has been deeper conflicts within different
governmental bodies than between the state and the environmental organizations.

Also, when it comes to the development of watercourses, we do see the
importance of the local community perspective. The development of watercourses entails an industrial and a local part. Technological development in the
beginning of the 1900s made large-scale industry dependent on hydroelectric
power. Alternatively, large-scale industry became possible once one understood how to generate hydroelectricity.[28] In this process, large dams were built
and waterfalls and rivers diminished to trickles or altogether disappeared into
pipelines. However, the development of watercourses also led to an intense
electrification of the country. Some municipalities and local entrepreneurs
benefited immensely from this, as did small-scale industry and the general
public (Sejersted, 1993, p. 177). Some of the wealthiest (per capita) municipalities in Norway are those whose revenues mainly stem from production of
hydroelectricity. Because ownership, due to the foresightedness of national
politicians, at least over time was returned to local municipalities, money has
remained in the local community.[29]

---

[27] The government-authorized wolf hunt reached the airwaves of CNN and other international television channels and did not really give Norway the kind of publicity that it had
anticipated.

[28] This was also linked to the production of fertilizers in part explored by the Norwegian
pioneers Christian Birkeland and Samuel Eyde and that led to the establishment of Norsk
Hydro 1905.

[29] Some of this industry has received state subsidies, and not all said municipalities have
become wealthy.

Finally and perhaps also most conspicuously and strongly related to the above discussion, the Norwegian brand of environmentalism, contrary to environmentalism in most other countries, excludes animal rights. The Rootes study also finds that animal rights is not always fully integrated into the overall environmental movement, especially in Britain. However, this partial absence stems mainly from the lack of coordinated action across different environmental fields. In the Norwegian case, exclusion of animal rights from environmentalism takes place on a more general and profound level. Most of the time when people think of environmentalism, animal rights are cognitively not included in the concept (see Chapters 5 and 7). With the exception of the organization NOAH—for animal rights and partly Greenpeace, both of which are part of our study, the official policy of the other 10 environmental organizations in our study is that protection of animals is not part of the definition of environmentalism. This result emerges quite clearly from our interviews with the organizational elites too.[30]

The absence of animal rights can, in part, be explained by the more overall pragmatic political culture and the rather weak urbanized understanding of nature within that culture. The reason for this being so, we believe, is that the local community perspective within Norwegian environmentalism entails that the protection of nature include the protection of human beings in close relations to nature. The protection of humans in nature is at least as important as the protection of nature itself. Thus, there is *not* a dichotomy between urbanity and humans on one side, and nature and animals on the other. The dichotomy distinguishes between, on the one hand, an urban life and, on the other hand, a local life in close relations to nature. This is the essence of the local community perspective in Norwegian environmentalism.

Two Anomalies Make a Unique Case

The state-friendly society makes the Scandinavian countries distinct compared to other countries. The local community perspective makes Norway distinct compared to other Scandinavian countries. We argue that the conjunction and interaction of these two anomalies, not the least through a tradition of strong local government, has made the Norwegian case of organized environmentalism unique in an international context.

The state-friendly society has moved Norwegian environmental organizations closer to the state structure and, to some extent, made the organizations dependent on the state. However, it has also moved the state closer to the organizations. Norwegian environmental organizations are relatively weak when defined in terms

---

[30] There is also a more moderate animal protection organization in Norway: the Norwegian Federation for Animal Protection, founded in 1859. This organization can, however, not be considered an environmental organization (see Bortne, Grendstad, Selle, & Strømsnes, 2001). See also the discussion on animal welfare in Chapter 7.

of membership numbers and local branches. This is very different from the picture within other parts of the voluntary sector in which single organizations can have several hundred thousand members.[31] Even so, the environmental organizations really matter. They have been an important part in a process that has transformed the political language and strongly influenced how governmental bodies operate within this important policy field. Whatever the policy area today, environmental concerns have to be taken into account.

A key factor here is that the state responds relatively quickly to demands of environmental organizations. The environmental organizations in the 1960s represented a new issue on which they quickly won public support. Thus, the state co-opted this issue rather fast, not the least because of the state's openness and the impact from civil society. The result is that the organizations, despite their weakness as membership organizations and their lack of ability to generate most of their finances on their own, have obtained political influence. Furthermore, it is very important to keep in mind that environmental issues and thinking did not only influence the state structure as such. Environmental issues and environmental considerations seeped into other voluntary organizations of the civil society more generally (e.g., welfare, culture, and leisure). However, because the other Scandinavian countries also have state-friendly societies with a strong voluntary sector, there must be something else that operates in Norway.

The Norwegian state is relatively new by European standards. Its final independence was obtained in 1905 after half a millennium under Danish (1397–1814) and Swedish (1814–1905) rule. It is a country without aristocracy, where cities are weak and small and where the national myth upholds individual independence, local community self-reliance, and egalitarianism. Norway, without a feudal tradition, has never fostered local elites strong enough to menace the state. No local elites have been able to veto policies and political aims. The state has never used its police or military forces to repress its citizens or hold them at gunpoint.[32] The level of societal violence is low and police, by default, carry out their duties unarmed. Citizens have for long trusted the state as a problem-solver and welfare-provider. Taken together, this has led to the growth of a strong and highly legitimate state in close contact with its citizens (see Chapter 8).

References to anything local are often used rhetorically to invoke what basic characteristics of Norwegian politics are and to identify roots of genuine Norwegian values. For example, in the heated debates on Norwegian membership in the European Community in 1972 and in the European Union in 1994, antagonists heralded Norwegian local government as a counterpoint to the ossified and opaque bureaucracy in Brussels. The battle cry was: "It is a long way to Oslo, but the road is even longer to Brussels." Studies of the two referendums on Europe showed that the more peripheral the area, the stronger the no vote. This local

---

[31] For further discussions on this discrepancy, see Selle (2000), Sivesind et al. (2002), as well as Wollebæk and Selle (2002a).
[32] There are minor exceptions—for instance, the Menstad confrontation in 1931.

perspective is accompanied by a large trust in the state institutions coupled with large discretionary powers in local politics.

The structure of the voluntary sector in Norway highlights the link between state-friendliness and a local orientation. To a very large extent, the same organizations have offices both at the central level as well as at the local level. Norway, to a lesser extent than most countries, developed a kind of dual organizational society (Wollebæk & Selle, 2002a). This lack of a dual organizational society entailed that members who were active at the local level of the organization had fairly unobstructed access to the central level of the organization.[33] The organizational elites, who were negotiating with the state, was, from the perspective of the local members, one of them. This structure was strengthened by the historically close relationship between central and local governments in a system of comprehensive local autonomy. The role of representative government and representative democracy has been very strong in Norwegian politics. Altogether, this is a system in which the combination of civil society and state relations operates very differently from what can be understood on the basis of the Dryzek study.[34] Furthermore, an emphasis on different environmental cultures—as the Rootes study does—would not have taken us very far in understanding the uniqueness of and the operation of Norwegian environmentalism.

Because of the state-friendliness and the local community orientation, the environmental movement continued to be pragmatic and moderate and does not hesitate to work closely with governmental bodies. When the organizations are in conflict with governmental bodies, which is not at all that uncommon, it does not seem to have any long-term consequences for the integration between government and organizations. With few exceptions, this way of life resulted in a rather moderate and nonfundamental type of environmental organization. As we will discuss in Part II, this is also strongly reflected in the attitudes and behaviors of the organizations' members. Hardly any of the environmental organizations developed a distinct green ideology. Also, none of the broad environmental organizations include animal liberation as part of environmentalism (see Chapter 7). Today, the Norwegian environmental movement is a rather pragmatic one. The local community orientation simply keeps this tendency in place. The local orientation is an essential part of the political culture in a political system that lacks

---

[33] For an interesting analysis on how the American voluntary sector also started out as rather integrative across geography, but later became much more dual with less grassroot influence on what is going on at the central level, see Skocpol (2003). This is a development with much in common with what has happened to the Norwegian voluntary sector over the last 20 years.

[34] In the Dryzek study, as in so many other studies of the state, you can get the impression that government is almost equal with state bureaucracy. The role of representative government and of the Parliament is played down. Both in general, and particularly in a Scandinavian context, we see no plausible arguments for such a view. For discussions on these points, see Tranvik and Selle (2005) as well as the final report from the "Power and Democracy" research project in Norway (Østerud et al., 2003).

strong subpolitical groups of any type (i.e., groups that for ideological reasons are unconnected to the government). Were the Norwegian society to have such groups at all, they would most likely be found within the religious rather than within the environmental field (Wollebæk et al., 2000). However, the political culture, so to speak, of state-friendliness and local community commitment entails a political system in which a vibrant civil society can be found.

State-friendliness also entails an open polity in which a green party failed to gain electoral success (see Chapter 8). This is partly due to the existing political parties being successful in preempting the environmental issue. However, the failure of a green party is also due to the openness of the polity in which political protests can enter the political system outside of party organizations. Political protest can enter the system through other civil society organizations that often have extensive contact directly with the public bureaucracy. As we have shown elsewhere (Bortne et al., 2002), this type of contact is very common.

Few Norwegian environmental organizations really look beyond national borders. Their orientation has a national and local focus. Only to a limited degree do they have contact with similar organizations in other countries. Although some organizations do move beyond and establish themselves outside of national borders, their move is not considered necessary for the environmental cause.[35] This is somewhat surprising when one considers that the environment as such knows no national borders and that most types of pollution, for instance, must be addressed more as an international than a national problem. This intranational position, we think, is a consequence of the organizations being locally oriented and perceiving the state as a friend. This intranational orientation is something other than the Rootes study's emphasis that so much of environmental protest behavior in Europe is national in character and related to political conjunctures in each and every country. In the Norwegian setting, we are talking about a strong cognitive or ideological orientation in which the mental energy is put mainly toward what is within your own borders. Let this be our frame of reference when we later in this book look at who the environmentalists are and how they think and behave.

---

[35] An exception is The Bellona Foundation, which opened offices in the United States, Brussels, and St. Petersburg. Because of government support, The Norwegian Society for the Conservation of Nature, Nature and Youth and Bellona are also present in north Russia. The international commitment is tempered by the fact that the pollution in north Russia can also severely affect Norway. Both WWF and Greenpeace are international organizations whose disproportionally weak representation in Norway is also a case in point. The Norwegian Society for the Conservation of Nature is the Norwegian member of Friends of the Earth. However, this has no strong influence on how the organization operates (Bortne et al., 2001; Strømsnes, 2001).

Chapter 3
The Organizational Setting: Early History and Later Developments

Introduction

In the Norwegian society, voluntary organizations have had a prominent position for more than a century. Between the late 1870s and the late 1960s was the period of the traditional social movements whose purpose was broad and to which members were important. Significant movements included the teetotalist movement, the farmers' movement, the lay-Christian movement, the "New Norwegian" linguistic movement, the labor movement, the sport movement, as well as important organizations within the social and welfare fields. These movements were organized independently of the state, but this did not exclude even strong and long-lasting cooperation with the state. The organizations had, however, a far more autonomous position than what is described as the "actively inclusive state" by Dryzek and colleagues (2003). At the same time, the social movements maintained close links to people living in the periphery of the country and they were crucial in the political and cultural transformation of the country, including the nation-building and democratization processes (Rokkan, 1970; Østerud et al., 2003).

During the 1960s, traditional movements' membership base declined, heralding changes in the way organizations operated. Traditional movements were also complemented with and challenged by organizations promoting leisure and special interests. These types of organization had slowly replaced their societal interest with those of their members. The 1960s is also the decade when "new politics" entered the stage. A new generation started to question central societal goals and consensual issues like economic growth, modernization, and technological development. Among the range of new issues, such as those advocated by the peace movement and the women's movement, environmentalism has been considered the most typical example of new politics (Dalton, 1994; Poguntke, 1993). Over the years, environmentalism permeated society and changed the language of politics, even if the number of members, or supporters, never became that high compared to other organizational fields. Today, public policy is incomplete if it fails to address environmental issues, individuals have the burden of proof if they are caught in environmental sins, and "environmental" has become prefixed with next to everything (even though the term might be deceptive for the actual content).

In the beginning of the 1980s, the role of the organizations' members was played down, parallel to the reduced importance of and need for voluntary work by members within many organizations. Organizations gave less priority to organizational democracy and member participation faded. In this period, the organizational field had become significantly different from that of earlier periods, and environmental organizations established after the middle of the 1980s were also different from those established earlier. We see the development of a new generation of organizations. In many ways, this change within the environmental movement is prototypical for a more general transformation of the voluntary sector (Wollebæk & Selle, 2002a). The organizations that were founded in this period have features that are as compatible with how organizations are described within the American tradition in the social movement theory as with the description that follows the European tradition. Whereas the European tradition stresses the normative/ideological and cognitive aspects of organizations, the American tradition places greater emphasis on the organizational and "entrepreneurial" aspects (McAdam, McCarthy, & Zald, 1996; Morris & Mueller, 1992).

Concurrently, newly formed organizations became more professionalized, specialized, and centralized. They adopted market logics and cooperated more frequently with market actors (Selle & Strømsnes, 1998). Staff became more professional and members became less important because organizational democracy quickly becomes an obstacle when organizational success is measured more in media exposure than in member participation. When new environmental organizations ask for funding, the argument is less the need to educate members in democracy and participation than to promote a vital societal cause through project support.[36] Some organizations now depend heavily on financial contribution from businesses and market actors. This emerging pattern is typical for a general change of the voluntary sector, but it has been most conspicuous among environmental organizations (Selle, 1996; Wollebæk & Selle, 2002a).

Although nature conservation organizations were established at the turn of the 20th century, the environmental *field* emerged considerably later. It is not until the late 1960s or early 1970s that the Norwegian society can count several environmental organizations.[37] This corresponds with the waves of environmentalism

---

[36] "Members" in new organizations are sometimes referred to as "supporters," "followers," "participants,"and so forth by the new organizations themselves. Beyond this chapter for simplicity we refer to all members of both old and new organizations simply as "members." We will distinguish between "members" and "supporters" in Chapter 9 when we discuss the changes among environmental organizations in the broader context of voluntary organizational change.

[37] Environmentalism became a permanent political issue in Norwegian politics at the end of the 1970s, when it ranked second on the average voters' account of important political issues (Aardal, 1993).

seen in most Western countries (Bramwell, 1994; Rootes, 2003). However, despite the prominent position of organizations in the Norwegian society, environmental organizations have failed to become broad popular movements. The membership rates have never been high, even though there have been more members in some periods than others. Shifting media attention, lack of popular causes, and policy contestations have contributed to the ups and downs of the environmental movement during the development of environmentalism as a political field. Even so, it is more than interesting to note that in a country with one of the most extensive voluntary sector there is, the number of organized environmentalist has been so low (Salamon & Sokolowski, 2004; Sivesind et al., 2002). Unlike the Dryzek study, we do not think this can be fully explained by governmental bodies adapting environmental concerns early, making mobilization more difficult and less necessary. We believe that this argument has some explanatory power. However, more important, we think, is the fact that it in general has become much more difficult for politically based organizations to get high membership figures. This is a key aspect of the transformation of the voluntary sector now going on and an important change with a great implication for democracy that we have seen coming over the last 20 years.[38]

Early History and the Classical Period: 1914–1963

The roots of environmentalism can be traced back to resource conservation in the late Middle Ages. Overpopulation in the period before the Black Death (1349–1351) forced Norwegian authorities to regulate hunting and fishing in order to prevent excessive harvesting (Berntsen, 1994). A more restrictive approach dates the start of environmentalism to when government resources for the first time were used to protect nature and environment (Jansen, 1989). This type of protection began in the 19th century, with the preservation of forests redefined as parks for public use. In contrast to the resource conservationism and wilderness preservationism that developed in the United States in the middle of the 19th century, incipient European environmentalism included conservation of nature for reasons of culture and civilization (Berntsen, 1994; Eckersley, 1992; Gundersen, 1991, 1996).[39] One of the first organizations founded in this field was The Norwegian Mountain Touring Association (DNT). Established in 1868, it is not only the oldest organization included in our study, but it is also a pioneer organization in Europe. However, DNT was not originally an attempt to promote the interests of nature itself. Rather, it was an organization for the "knee breech nobility" to go fishing, hunting, and mountain touring in the Norwegian mountain

---

[38] For an extensive discussion of why we think that is so, see Tranvik and Selle (2003, 2005).

[39] For an introduction to the international and historical growth of environmentalism, see Bramwell (1989), McCormick (1995), and Pepper (1996).

wilds (Berntsen, 1994). Such use of nature was inspired by British conservationism and the first tourists were indeed British aristocrats. This was the beginning
of mountain tourism in Norway.[40]

Still, DNT's main purpose is to organize outdoor recreation activities by mark-
ing and maintaining paths and trails and to operate tourist cabins. This is the
largest organization included in our study, with more than 200000 members and
50 local branch offices.[41] Although the organization always has been engaged in
conservation issues, its activities lie on the borderline of "environmentalism."
When environmentalism became a political issue, the organization included protection of the natural environment in its formal bylaws.

The first to advocate protection of Norwegian nature were scientists and natural historians. Particularly, certain artifacts, objects, or species were prime targets
of preservation (Jansen & Mydske, 1998). The National Association for Nature
Conservation ("Landsforeningen for Naturfredning," LfN), the predecessor of
today's The Norwegian Society for the Conservation of Nature, was founded in
1914. This organization was mostly a club for intellectuals and scientists of the
upper social classes and it failed to attract popular support. In the years preceding World War II, it counted approximately 1000 members (Berntsen, 1994;
Gundersen, 1991).

The scientific orientation and the recruitment from within the scientific
milieu was considerably stronger in the environmental field compared to most
other fields of the voluntary sector. The organizations were not for the common
people (Jansen, 1989). Especially this is the case when we look historically at
the environmental movement, but also today, this scientific orientation is
something that characterizes the movement and that has consequences for
mobilization and support. The scientific and upper-class background is also
important to explain why the membership rates in most of the organizations are
still relatively low.[42]

The period between 1914 and 1963 has been labeled "the classical period" in
Norwegian environmentalism, in which The National Association for Nature
Conservation set the public agenda for conservation of nature (Gundersen, 1996).

---

[40] Even though this kind of mountain tourism was for the upper class, simplicity and moderation have always been main values (Richardson, 1994). These are, at the same time,
essential values in the Norwegian culture, which is linked to what we call the local community perspective.
[41] Where available, we report the 2005 membership figures primarily found at the organizations' websites (see Appendix B), as well as the number of members at the time of our
survey. In 1996, DNT reported 182,000 members (Strømsnes & Selle, 1996).
[42] Also in the welfare field, the scientific milieu was strong within the voluntary sector,
including doctors and other professions. However, here the membership figures were
extremely high (Berven & Selle, 2001). We think this partly has to do with that questions
of welfare at the time went deeper than what environmental questions have done so far.
However, it is also related to the transformation of modern politics in which you can have
great political influence without high membership figures (especially if given the expression of being scientifically based).

Nature became cognitively important in the growing Norwegian nationalism from the mid-19th century. Norway's independence from Sweden in 1905 did not temper such nationalism, nor the importance of the local community perspective, as we will see later.

The independence from Sweden coalesces roughly with the start of the "classical period." Mountains and waterfalls were symbols of the new nation and were distinct for the Norwegian nature (Gundersen, 1991). In this period, activists proposed to set aside distinct areas as national parks because they were found to be important for the development of a Norwegian nation and a Norwegian national culture. However, most of the proposals were never implemented.[43] The first national park in Norway, Rondane National Park, was not established until 1962, 60 years after the first proposal in 1902 (Berntsen, 1994) and 90 years after the world's first national park was established in Yellowstone in the United States.

Nature conservation soon became synonymous with national characteristics. This is a school of German origin where there already was a tradition for the preservation of cultural memorials ("Kulturdenkmal"). In the beginning of the 20th century, it also became important to preserve natural characteristics ("Naturdenkmal"). These natural characteristics were also supposed to be *national* in that they represented the nature of the country. The preservation of forests, lakes, islets, mountains, and mountainous territories in small areas was the aim of this type of nature conservation (Berntsen, 1994). However, the Norwegian understanding of nature soon got a lesser romantic content than the German one (Witoszek, 1997), at the same time as Norwegians nourished a pragmatism heavily influenced by the local community perspective.

The first years of the new century became dominated by struggles over watercourse development licenses. Should waterfalls be developed and used for industrial purposes and modernization ends, or should they be preserved for their national and aesthetic value? This struggle was inseparable from the pride of being a newly independent nation. The 1905 independence from Sweden would be reversed if one gave way to foreign ownership of watercourses. It would be equivalent to revoking the newly achieved independence (Gundersen, 1991).[44]

The Norwegian state assumed an active role on issues of nature conservation when the parliament passed the 1910 Nature Conservation Act in which responsibility for nature conservation policy was handed to the Ministry of Church and Education. From 1910 to 1965, the ministry had two officers in charge of such issues. However, the protection of nature was almost a nonissue in this period. Economic growth with strong national control won at the expense of conservation

---

[43] This dualism—the importance of nature as mythology, but nevertheless politically neglected—is not easily explained. One reason is perhaps that it was hard to conceive that human activity could in fact harm the environment of such a vast wilderness.

[44] The same kinds of argument have been used in the EC and EU debates and subsequent referendums in 1972 and 1994, respectively—both of which resulted in a "no" to membership.

of nature. This might be understandable when we take into consideration that Norway then was a poor country in need of jobs, development, and modernization. Especially the years during the post–World War II reconstruction were a period of heavy industrialization and economic growth. Any public concern to environmental issues was the exception, and isolated cases had no consequences for everyday politics (Jansen, 1989).

In 1934, for the first time involving an environmental organization, the state initiated an annual financial support to The National Association for Nature Conservation. This support has continued ever since. In 1939, the state support amounted to 25% of the organization's income (Berntsen, 1994; Gundersen, 1996). Sixty years later, its successor, The Norwegian Society for the Conservation of Nature, received 23% of its income as basic grants from the Ministry of the Environment, but it also received economic support from other ministries and from other parts of the environmental administration—all in all, between 40% and 50% of its income (Bortne et al., 2001; 2002). This is somewhat more than for the voluntary sector in general (Sivesind et al., 2002), but far from enough to conclude, like the Dryzek study does, that for financial reasons, environmental organizations are arms of the state (see Chapter 2).

Although the practice of state financial support to other types of voluntary organization was not uncommon, the financial support to The National Association for Nature Conservation starting in 1934 shows that the close ties between state and environmental organizations were initiated and established early.[45] The financial support is also due to the organization's scientific base and to the fact that there did not exist a clear separation between bureaucracy and science. In many cases, we might also find the same persons on both sides of the table—both as members of organizations and as bureaucrats administrating economic support to the organizations (Bortne et al., 2002). There have been several examples of this kind of mix within the environmental field, but it is also something that, for instance, can be seen within the sectors of welfare and sports (Selle, 1998).

From 1963 and onward, one can identify the modern period of organized environmentalism. Prior to 1963, environmentalism was something for the very few, often based on both science and class. The period after 1963, which is characterized by increasingly close relations between environmental organizations and the state, can be divided into five more or less distinct subperiods.[46]

---

[45] For an analysis of the relationship between government and voluntary organizations in Norway from the mid-1800s to the mid-1900s, see Kuhnle and Selle (1992a). There exists no golden age of a strong and independent voluntary sector in opposition to the state. Close cooperation and financial support was common from the early days. This cooperation is deeply embedded in history (see also Tranvik & Selle, 2003).

[46] The periods are sufficiently distinct as to their substance, but insufficiently distinct as to their precise temporal demarcation. We therefore concede to a slight overlap of the years defining their start and end.

The Establishment: 1963–1969

During the 1960s, environmentalism as a *policy area* seeped into routine politics and the new political field became populated with influential thinkers and activists, among them most notably Arne Næss.[47] In addition to participating in the public debate and legitimizing environmentalism academically, thinkers and activists also participated in civil disobedience and in confrontations with the authorities.

In 1963, The National Association for Nature Conservation was reorganized and renamed The Norwegian Society for the Conservation of Nature (NNV). Whereas the former organization had been elitist and scientific, the latter became democratic and member-based. Here we define an organization to be democratic when it has ordinary *members* and not only supporters, and when members have, at least formally, the possibility, preferably through the organization's local branches, to *influence* the central level in the organization, ultimately leading to complete change of leadership. The Norwegian Society for the Conservation of Nature worked against watercourse development and in favor of building national parks. However, the new organization still favored economic growth and environmental protection. The Norwegian Society for the Conservation of Nature is the Norwegian member of Friends of the Earth. Its approximate 16,800 members in 2004 (a drop from 40,000 members in 1991 and 28,000 in 1996) are organized in about 170 local branches. If the Norwegian environmental movement has had a core organization, this is the one. In recent years, it has been severely weakened like so many other organizations building upon the traditional organizational model (Wollebæk & Selle, 2002a), but the organization is still among the largest environmental organizations in Norway and it is represented in a number of government boards and committees.

Nature conservation issues were reorganized at the ministry level in 1965. Ministerial responsibility for nature conservation was given more autonomy and moved to the Ministry of Labor and Local Government. Here a subdivision for Outdoor Life and Nature Conservation was established. The environmental policy of the new institution combined economic growth and environmental protection. However, the subdivision was hampered in its work because of strong conflicts with The Norwegian Society for the Conservation of Nature. These conflicts were both of personal character and conflicts over issues, among them the subdivision's priority of outdoor recreation at the expense of nature conservation (Gundersen, 1996).

In 1967, two former biologically concerned youth organizations merged into Nature and Youth (NU). This is the first organization to unite youth and the struggle for environmentalism. Youth is secured by expelling members when they turn 25 years old. In 1968, Nature and Youth became the autonomous youth

---

[47] Main publications are by Næss (1973), Kvaløy Setreng (Kvaløy, 1972, 1973), Sætra (1973), and Dammann (1979).

organization of The Norwegian Society for the Conservation of Nature. Influenced by the "ecosophy" and "deep ecology" formulated by Næss (1973) and Kvaløy Setreng (Kvaløy, 1972), Nature and Youth aims for a more holistic understanding of nature and environmentalism. It maintains a local orientation, evidenced by its alliance with the Smallholders' Union, implying, for example, that the organization recruits urban youth for summer jobs at farms in order to give them experience with practical and productive work ("matnyttig arbeid"; Haltbrekken, 1996). This alliance illustrates the importance of the local community perspective. The result is an environmental youth organization with a very pragmatic attitude rooted in strong ideological convictions (see Chapter 7). This is also the case when it comes to questions about whale and seal hunting. The organization is in favor of both whaling and sealing as long as there are "enough animals" (Haltbrekken, 1996). Nature and Youth is a democratic organization and it operates as a member organization. It has roughly 5000 members and 80 local branches (down from 6000 members and 130 local branches in 1996).

During the 1960s, environmental concern became a much discussed topic at the universities, and especially at the University of Oslo. In 1969, Environmental Co-operation Groups were established and organized at the Department of Philosophy under the leadership of the philosopher-turned-ecological farmer Sigmund Kvaløy Setreng. Although members of the groups most of the time discussed philosophy, their aim was "Gandhistic direct action" (meaning that actions were nonviolent and announced before the event) (Kvaløy Setreng, 1996).[48] Different groups were given different tasks, such as direct actions, making maps, organizing exhibitions, and exploring biological issues. It was argued that small groups were more efficient than large ones. The "small is beautiful ideology" was strong within these groups. However, in spite of its success, the loose organization faced problems of organization and overall policies (Gundersen, 1996). Attempts from roving Marxist groups to take over some of the environmental groups in 1975 and 1976 were partly successful (Knutsen, 1997).

Taken together, the environmental field in this period was marked by influential thinkers and a high level of activism. It prepared the ground for an increasing politization in the 1970s and for the establishment of more specialized organizations. However, even if in deep conflict with government, the organizations did not turn away from the state, but mobilized to improve or even transform state action.

Breakthrough and Politization: 1969–1981

The goal of The European Nature Conservation Year in 1970 was to advance greater environmental concern. In Norway, Prime Minister at the time, Per Borten, ceremoniously chaired the national committee. This environmental campaign was

---

[48] Information in this paragraph is in part based on a long letter from Sigmund Kvaløy Setreng to the authors.

deemed a success and the environmental movement gained momentum (Gundersen, 1996). In Norway, the years between 1970 and 1975 have been called "The Golden Age of Environmentalism" (Berntsen, 1994). The label is credited to the establishment of the Ministry of the Environment in 1972 (which was one of the first ministries in this field in the world), new acts on nature conservation and water pollution being passed in 1970, increased environmental concern in the general public, and environmental issues being broadly covered in the media. During these years, the Norwegian environmental movement was claimed to be the strongest in Europe (Berntsen, 1994). This fits well in with the Dryzek study's emphasis upon Norway in this period being the greenest of states.

Following the establishment of the Ministry of the Environment, "nature conservation" (*naturvern*) was redefined as "environmentalism" (*miljøvern*), with greater emphasis on pollution. This conceptual upgrade was a sign of the increased importance of the field as to which problems were addressed and how policy was implemented. The new ministry was given increased authority both within the environmental policy field as well as in coordination across several policy fields. The dilemma between economic growth and environmental protection was Salomonically solved by the policy of "growth with protection" (Jansen, 1989, p. 248). The ministry initiated new policy, revised laws, as well as made new laws. The Ministry of the Environment "was particularly active in the field of pollution abatement in which it mainly pursued *the Polluter Pays Principle*" (Jansen & Mydske, 1998, p. 183).

Even today, the Ministry of the Environment plays a very important role within the environmental field in Norway. The fact that the Ministry was established already in 1972 implies that this field was co-opted early by the state. This is something that has consequences for the movement later. As Dryzek and colleagues (2003) pointed out, inclusion almost invariably means moderation. Organizations are not that different in ideology and policy interest than governmental bodies themselves.[49] However, as argued in Chapter 2, rather than seeing the organizations being pressed to become moderate, we see them as being moderate from the start. It is this symmetry that makes cooperation come so naturally for both parts. However, the moderation argument should not be taken too far. Being in conflict with government bodies does not at all automatically mean exclusion from government bodies.

Ironically, in Norway, The European Nature Conservation Year coincided with the environmental battle in which the government planned to build a hydroelectric power plant supplied by the water of the Mardøla River. Environmentalists mobilized against the project and attracted unprecedented media coverage. After a series of confrontations in August 1970, the environmentalists conceded defeat. The symbolic value of the Mardøla incident stems from it being the first true confrontation on environmental protection versus economic growth. A picture

---

[49] Political inclusion might place strains on movement groups in a way that undermines participatory and decentralized structures (Dryzek et al., 2003, p. 83).

showing the philosopher Arne Næss being carried away from the scene by the police epitomized the incident.

Although environmentalists can count some victories on watercourse development in this period (e.g., Veig and Dagali), they lost another symbolic battle in the beginning of the 1980s when the water in the Alta River in the Sami territories in northern Norway was used for production of hydroelectric power.[50] Although the environmental movement lost the 1970 Mardøla battle, the confrontation imbued the movement with strength and optimism. Inversely, the movement could in a sense claim a small victory in that the fight over the Alta River led both to reduced watercourse development and, not the least, increased recognition for the rights of the Sami people.[51] However, here, the movement was unable to capitalize on its gains. This paradox, that members are mobilized when the movement loses the fight against the developers but not when they have success in their fight, is not easily explained even if related to the overall standing of environmentalism in the period (see discussion later in this chapter).

During this period, new and more specialized organizations were founded. First founded internationally in 1961, the Norwegian branch of World Wide Fund for Nature (WWF) was founded in 1970.[52] The organization specializes in wildlife protection and the survival of endangered species. It works first and foremost on global environmental issues. Its main goal is to preservere the diversity of nature and to ensure a sustainable use of natural resources. WWF is important primarily due to its international standing, referring to itself as "the world's largest nature conservation organization." In Norway, WWF has approximately 5000 members (down from 6000 in 1996). Despite it being established in the "democratic" period (see discussion later in this chapter) but mainly due to it being an international organization, WWF has not built a democratic organizational structure with regular membership through local branches. WWF's Norwegian branch has simply been organized in the same way as the international organization and was, up to March 2000, organized as a foundation. After this date, it was established as a democratic organization in which members gained greater influence. One of the reasons for this change was to be eligible for state funding (WWF, 2000) and the organization has been included in the state's national budget since 2001. This democratic concession is, however, mostly formal and carries little substance. The organization is still part of the international

---

[50] Many years later, Gro Harlem Brundtland and Odvar Nordli, minster of the environment and prime minister, respectively, during the Alta watercourse development debate, admitted publicly that the development of the Alta River was a mistake.

[51] Soon after the fight over the Alta River ended, the Norwegian government gave up its strong integration policy and allowed the Sami community to develop its own public space. In 1989, the Samis got their own Parliament with a certain autonomy, an autonomy that has increased ever since. For the development of a new Sami public space, see Bjerkli and Selle (2003).

[52] Until 1986, World Wide Fund for Nature was named World Wildlife Fund (Guldberg & Schandy, 1996, p. 150).

foundation, and the relationships among the old Norwegian organization, the new member organization, and the international organization remain unclear. Also due to its type of organization at the time of our survey, WWF will, in the remaining parts of this book, be considered a nondemocratic organization.

In 1971, The Norwegian Organization for Ecological Agriculture (NØLL) was established. The organization sought to unite consumers and producers for the cause of ecological or organic agricultural products. The focus on ecological products was extraordinary in a Norwegian setting, and the organization was also unique because it targeted as members both consumers and producers (i.e., farmers). The organization was democratic in structure and had about 1300 members and 21 local branch associations in 2000 (as in 1996). In 2001, NØLL contributed to establish a new organization—Oikos—consisting of a total of three organizations, all of which advance ecological products. This is a new kind of interest within Norwegian environmentalism. However, the spread of ecological production and consumption in Norway has not been very comprehensive when compared to the situation in most other Western European countries—for example Denmark, the United Kingdom, and Germany (see Terragni & Kjærnes, 2005). The new Oikos organization has about 1600 members.

Another organization established in the 1970s and which promotes an alternative life style is The Future in Our Hands (FIVH). It was set up in 1974 and consists of more than 20,000 members and 18 local branches (16,000 members and 30 branches in 1996). This is, then, a rather large organization in a Norwegian context. In addition to its commitment to environmental conservation and quality-of-life questions, the organization keeps the cause of the developing countries on its agenda. This mix gives the organization its special character. Through its global agenda, FIVH claims to comprehend environmental problems and societal challenges better than other organizations. It is democratically built, with its members represented at a national congress and in an executive board, but nevertheless rather centralized in practice (Strømsnes, 2001).

Taken together, the 1970s were successful but also politicized years for the environmental movement. Environmentalism became institutionalized in Norwegian politics, especially with the establishment of the Ministry of the Environment. Electoral studies also show that environmentalism became a salient issue in politics in this period. In the 1977 Storting election, the issues of growth, energy, and environment were the second most important, only surpassed by the issue of the role and size of the government (Valen, 1992). However, during the late 1970s, conflicts and disagreements took place within and between environmental organizations on how to solve environmental issues politically and how to address environmental problems in the context of the establishment of the Ministry of the Environment. Vested political interests, such as the Labour Party and the Conservative Party, fought successfully against the environmental movement over energy politics, in that the two parties advocated the development of new oil fields and watercourses. Thus, it should come as no surprise that these parties are not doing well among organized environmentalists (see Chapters 5 and 6). However, since the discovery of oil outside the coast of Norway in 1969, the development of the Norwegian oil fields in the North Sea has also had an effect on the political will to pursue nuclear energy and watercourse development.[53] The politicization in the 1970s led to environmental organizations finding themselves both in cooperation and confrontation with state authorities. On balance however, the environmental movement enters the 1980s with less hope of matching the success of the 1970s.

Recession: 1981–1988

Similar to other Western countries in the 1980s, Norway experienced conservative-led governments speaking the gospel of deregulation and liberalization. The environmental movement fought, in many ways, an uphill battle. Environmental ideology lost attraction and the movement declined (Gundersen, 1996). However, the one environmental issue that received increased attention during the 1980s was pollution, which should prove powerful in combination with the increasing direct-action-oriented behavior among organized environmentalists (Berntsen, 1994). In addition to processes inside the organizations and an increasing need for faster and more efficient solutions, the liberal political context paved the way for a new type of organization. We see in this period a growth in the kind of organizations that Jordan and Maloney (1997) call "protest businesses." Compared to previous organizations, the new organizations were centralized and nondemocratic. Being untied to a membership-based organizational democracy, activists could strike against polluting factories and companies without being hampered with the internal bureaucracy of its organizations. Concurrently, these organizations developed a new definition of efficiency. In the 1970s, efficiency implied a democratic debate within the organization, thus reaching a consensus through mutual understanding and a common ideological platform. Now internal democracy was subdued and efficiency meant the ability to act quickly and decisively when pollution had to be stopped and media attention was demanded. Organizations established in the 1980s and 1990s are often found to be more professionalized, more centralized, more action-oriented, and more focused on solutions than were the organizations established earlier (Strømsnes, 2001; Strømsnes & Selle, 1996; Tranvik & Selle, 2003). This organizational development is part of a broader process that is found within the voluntary sector in general (Wollebæk & Selle, 2002a). However,

---

[53] Environmental resistance against development of watercourses cannot be taken for granted. In Sweden for instance, many environmentalists favored such development because to them the energy alternative was nuclear power plants. During a brief period in the mid-1970s in Norway, some environmentalists argued that nuclear energy should be favored in order to protect watercourses. In general, however, resistance against nuclear power was strong. The 1979 nuclear accident in Harrisburg, Pennsylvania, USA, effectively put an end to the idea of nuclear power in Norway (Berntsen, 1994, p. 289).

the process can be seen most clearly within the environmental field (see Chapter 9).

Two prominent organizations were founded in this period, both of which made pollution one of their most important issues (Gundersen, 1996). First, The Bellona Foundation (MB) originated from Nature and Youth in 1986 as a high profile, direct-action organization. MB is not a democratic member organization. It includes approximately 1000 supporting members (3000 in 1996). Much of the organization's incomes are grants from businesses in addition to project support from the government. MB has been the most visible of the new organizations since the 1980s and has been decisive for developing new tools in the environmental battle.

Second, the Norwegian branch of Greenpeace was founded in 1988. Greenpeace Norway (GN) had 200 supporting members in 2000 (down from approximately 750 members in 1996). This makes it very small compared to the dominance obtained by Greenpeace International, which was founded in 1971.[54] GN later merged with its sister organizations in Denmark, Finland, and Sweden, with a new headquarters established in Stockholm. This Nordic organization has about 105,000 members and highlights the extremely weak position that Greenpeace has in Norway, especially when compared to neighboring Sweden.

One of the reasons for the marginal position of GN is its domestic wrong-footed policy against Norwegian whaling. Another reason is that MB and its mediagenic leader, Frederic Hauge, preempted GN's potential for popular support in that MB was founded 2 years before GN. However, an even more important reason, we think, is that GN is not in harmony with the local community perspective. GN's fight against Norwegian whaling and sealing is just one example of the fact that the organization does not understand, or is unwilling to accept, that in Norway, environmentalism is important only to the extent that it does not threaten the survival of local communities (see Chapter 7).

Both MB and GN keep environmental experts in their staffs. Both attract media attention through direct action. Also, despite their differences in origin and organizational competition, the two organizations have a record of close cooperation in actions against polluters (Nilsen, 1996; Strømsnes 2001). An important distinction between the two is that members are financially less important in MB than in GN because GN does not receive public or business funding due to its strict policy of independence. This also implies that the organization is not integrated in the state sphere to the same extent as other environmental organizations in Norway. Consequently, personal networks decisive for public funding and legitimacy are lacking (see also Chapter 8).

---

[54] Greenpeace is, however, in trouble in several countries. In the United States, the decline in membership has been dramatic.

Prosperity: 1988–1993

Following the materialist and liberal political wave in the early 1980s, the Norwegian environmental movement resumed its strength, although counted in membership numbers it has never been a strong movement. A number of events highlight the context of this period of prosperity. Most important, perhaps, was the Chernobyl nuclear accident in April 1986. The nuclear power plant was situated north of the city of Kiev, Ukraine, in the former Soviet Union. The explosion of a reactor, and its continual burning until October, was the worst accident in the history of nuclear power. Radioactivity soon spread by the wind and reached large parts of Europe, leading to contamination of farmland and evacuation of thousands of people. Neighboring Norway was strongly affected and consequences of the accident are still observed.

Environmental issues compiled during the 1980s included the following: the spread of algae in seawaters, which was also very dramatic in Norway because of its long coastline; general pollution of the environment; depletion of the ozone layer; the greenhouse effect and global warming. In 1987, the UN World Commission on Environment and Development, chaired by the Norwegian prime minister Gro Harlem Brundtland, issued *Our Common Future* (Brundtland et al., 1987). The report coined "sustainable development" as "development that meets the needs of the present without compromising the ability of future generations to meet their own needs" (Brundtland et al., 1987:43). Unable to escape the dilemma, the commission suggested "growth with protection" even at the global level.

In June 1992, the United Nations Conference on Environment and Development in Rio de Janeiro, Brazil, also known as the Earth Summit, was the largest gathering of world leaders in history, with 117 heads of state and representatives of 178 nations attending. The final declaration laid down 27 broad, nonbinding principles for sound environmental development. Later, politicians found difficulties in implementing these principles and the global environmental strategies continue to divide industrialized from developing countries.

At a more local level, electoral research showed that the environmental issue was the second most important issue to voters in the 1989 Storting election (Aardal, 1993). Also in 1992, the Norwegian Parliament changed the Constitution. The protection of the environment and the citizens' right to a healthy environment was integrated in the text.[55] This is a very important institutional expression of the new role of environmentalism in politics. Without necessarily replacing nationalism with environmentalism, it is fair to say that, during this period, the "green flag was hoisted all the way to the top" (Berntsen, 1994, p. 9). This illustrates that even though the movement has never been strong counted in members, it has had a strong effect on politics. This is partly due to

---

[55] Paragraph 110b in the Norwegian Constitution, added by constitutional provision no. 463, 19 June 1992.

the changing role of politics, where strong public support no longer is a necessary supposition to gain political influence.[56]

Some of the organizations formed in this period carry on the demand for efficiency generated in the 1980s, whereas others pick up the environmental resurgence and euphora of the time. Yet the common denominator of the new organizations is that they heed less to democratic ideals and that they promote direct action. NOAH—for animal rights (NOAH) was founded in 1989. This is a small, radical, and active organization. It refers to itself as an interest organization for animals and seeks to restrict industrial animal husbandry and opposes the fur industry and the use of animals for research and entertainment purposes. The organization is registered as a foundation and does not count as a democratic organization. It consists of approximately 2000 subscribing members, of which roughly 1 in 3 is considered active (up from 700 supporting members in 1996). NOAH has 10 local branches that are represented at the national congress, but they are limited to voting on quite central amendments only (Strømsnes, 2001).

Another new and interesting organization of this period is Green Warriors of Norway (NMF). It is a splinter group set up after major conflicts in The Norwegian Society for the Conservation of Nature in 1993. It aims to become a nationwide organization but has so far limited most of the activities of its approximately 1500 members (up from 1000 in 1996) to the southwest coast city of Bergen, in addition to offices in two other cities (Oslo and Tromsø). In many ways, the organization is unlike MB in that it is more focused on regional and local issues. Nevertheless, NMF is frequently in the public eye. It is engaged in extensive consulting for trade, industry, and the public sector. Its structure is centralized around its leader, Kurt Oddekalv, who represents an action-oriented and media-focused type of environmentalism. Due to his rather dominant leadership style, Oddekalv might be considered an innovator of the Norwegian environmentalist movement. The organization is divided into ordinary members and active members, but only activists and organizational employees are allowed to vote in elections. Thus, it falls short of being classified as democratic.

In this period, we see the birth of a small and political interesting organization. Women–Environment–Development (KM) was founded in 1991 when it branched out from The Norwegian Housewives' Association (NH) (later to be renamed into The Norwegian Woman and Family Association). The defectors argued that NH was doing too little to protect the environment. At the time of the survey, KM consisted of not more than 140 members and is by far the smallest of the 12 organizations in this study.[57] KM attracts special interest due to its merger of women and environment, both of which are key issues of new politics. The organization's moderate ecofeminist profile, a novelty to Norwegian

---

[56] For a discussion of this important change in how modern politics works, see Tranvik and Selle (2003).

[57] Shortly after our survey, KM went into a sleeping stage. Later, it dissolved itself as an organization.

environmentalism at the time of its foundation (Gundersen, 1996; Strømsnes 2001), emphasizes that the experiences of being a woman is conducive to them giving greater priority to environmental protection than do men. This kind of organized ecofeminism is extremely rare in a Norwegian context and is the main reason for including such a small organization in our study (see Chapters 5 and 6).

Also The Environmental Home Guard has a special focus on green or environmental-friendly consumption similar to the KM.[58] However, the ideological setting between the two organizations is very different. The goal of The Environmental Home Guard is to promote responsible consumerism and consumer environmentalism. The Environmental Home Guard, which refers to itself as a working bee, is an umbrella organization established in 1991, covering individuals, companies, associations, and other organizations. It differs from the other organizations in that 15 voluntary organizations are gathered under its umbrella, among them we find three organizations that are part of this study: The Norwegian Society for the Conservation of Nature, The Future in Our Hands and The Norwegian Mountain Touring Association. The organization also recruits individual "participants," who are committed to nothing else but in their private lives to make an effort to behave environmentally friendly (e.g., recycle or compost household waste, avoid disposable products or products without an environmental seal of approval, repair broken articles rather than buying new ones, and use unbleached paper) (Endal, 1996). The organization has grown rapidly to more than 100,000 participants (up from 75,000 in 1996). However, because there are no mechanisms for withdrawal of memberships, figures will continue to increase. Consequently, this is a very different kind of organization than a traditional voluntary organization, including consumer orientation and individual responsibility, but without ordinary membership (see Chapters 5 and 6). It is, so far, a "deviant" organizational model both within the environmental movement and within the voluntary sector at large (Wollebæk & Selle, 2002a).

The Environmental Home Guard receives the largest sum of basic state grants of all the environmental organizations. It is a paradox given the weight the Ministry of the Environment puts on the classical democratic way of organizing when deciding, for example, which organizations to support financially (Bortne et al., 2002). The Environmental Home Guard is typical for an organizational development that is backed by the Ministry of the Environment. In fact, the Ministry of the Environment was instrumental in the founding of this organization. The ministry wants fewer organizations to cooperate with and, therefore, supports the establishment of cooperation forums and umbrella organizations, where a variety of organizations speaks with one voice when communicating with

---

[58] Later, The Environmental Home Guard was renamed Green Living Norway. In the following, we will use the name The Environmental Home Guard, which was in use at the time of our survey in 1995.

the authorities (Bortne et al., 2002).[59] However, often umbrella organizations, which cover a range of different organizations, will experience ideological and political conflicts that flourish below the surface. Because it can be difficult to argue clearly and strongly when you have to justify the arguments in front of several other organizations, this might also contribute to a development toward more moderate and pragmatic organizations.

Stagnation and Increased Contestation in the 1990s

The period from approximately 1993 on is one of stagnation for the environmental movement. This is in congruence with the development in the voluntary sector at large and also fits with the increased institutionalization of the environmental movement in general, underlined by the Rootes study. There is no increase in the number of organizations or in membership. However, a transformation of the voluntary sector is taking place. We see an increase in organizations similar in structure to the new environmental organizations of the mid-1980s while old, democratic, and more hierarchically based organizations are losing ground (Tranvik & Selle, 2003). Within the environmental field, apart from less than a handful of ad hoc organizations and networks, this period is conspicuous by the absence of any new organizations entering the environmental field.[60] Some network organizations and umbrella organizations have been established, where even the Ministry of the Environment has been involved, but traditional organizations have not been set up in this period. It is also a problem for the environmental movement that organizations that traditionally have been on the outside of the environmental movement incorporate environmentalism into its bylaws, thereby expanding, if not diluting the environmental field. This makes the environmental field less clear and exclusive, but it fits in with how governmental environmental bodies operate (Bortne et al., 2002).

Second, and related, no environmental issue in this period has had the potential to mobilize public interest and media attention in the same way as one has observed during earlier periods. Not that global climate change and depletion of the ozone layer is unimportant. However, such remote issues impinge less upon

---

[59] Perhaps we will see increasingly more of such direct involvement as part of New Public Managements' emphasis on efficiency and cost-effectiveness at the expense of organizational autonomy (Tranvik & Selle, 2003).

[60] One exception is the international organization Attac. Attac (Association pour une Taxation des Transactions financières pour l'Aide aux Citoyens) is an international organization that fights market-driven globalization. Attac can also be seen as a kind of environmental organization. In Norway, this organization has chosen a traditional, democratic, organizational form, including local branches and ordinary membership (see Jørgensen, 2001), which is an organizational model that is invariably less common among newly established organizations.

people's daily lives as do contaminated drinking water, dead fish, and the disappearance of a majestic waterfall. Once key environmental issues fade from one's attention or are "solved" and thereby removed from the agenda, a certain lack of focus and direction can be observed from actors who thrive on them.

This is also a period of increased contestation. Environmental actors and organizations have been accused of mixing facts and fiction. Some claimed that environmental risks might be unfounded or exaggerated. For instance, the Danish statistician Bjørn Lomborg, once a green activist, now holds a rather optimistic view and claims that there are plenty of energy resources and raw materials for the future. New oil fields are discovered all the time, and some energy resources, such as solar and wind energy, are renewable (Lomborg, 1998, 2001). Lomborg, among others, criticized the environmental movement for misinforming people about the state of the Earth using myths and fears that cannot be properly documented.[61] Lomborg further claimed that many people fail to understand that some environmental organizations, like other voluntary organizations and nongovernment organizations, are in the business of staying in business. And like some do, for instance, one can argue that Western values like individual freedom, which has brought us the affluence that permits the environmental agitators in the first place, might be compromised were the environmentalists' precautionary principle to be the sole guide to public policy (Rayner & Malone, 1998b; Wildavsky, 1995).

Since the end of the 1980s, Norway's official environmental position has been the politically correct policy of *sustainable development*. The challenge however, as Jansen and Mydske (1998) pointed out, is, when both cannot be accomplished, which of the two should be given priority. As long as sustainable development was linked to "economic growth with conservation," conflicts seem not to abound. However, when sustainable development became linked to "economic growth with preservation," mainstream policy gave way to growth when conflicts abounded (Bortne et al., 2001, 2002; Jansen & Mydske, 1998). The general observation, however, is that government's overall policy has incorporated and conventionalized environmental policy in a more and more institutionalized policy field. At the end of the almost 40 odd year period, environmental politics has become an integral part of politics in general (Jansen & Mydske, 1998).

---

[61] Kurt Oddekalv, the leader of Green Warriors of Norway, in a response to Lomborg, said that "Such [views] can only come from a person who does not understand ecology and environmentalism" (in Bergens Tidende, 1999, p. 5, our translation). Steinar Lem, the information officer in The Future in Our Hands, has called Lomborg "a David Irving in the environmental debate" (Lem, 2000, our translation). Lomborg has also been met with heavy criticism from within the scientific community, with negative reviews in journals like *Nature* and *Science* (e.g., Grubb, 2001; Pimm & Harvey, 2001). There were also lively discussions in Denmark after Lomborg was appointed director of Denmark's Environmental Assessment Institute in February 2002, a position he had until July 2004. In 2003, he was accused of dishonesty and lack of good scientific practice by the Danish Committee on Scientific Dishonesty, but he was later acquitted (see also http://www.lomborg.com).

The 12 organizations we have presented, described, and accounted for fall into a pattern that can be presented in a more parsimonious classification, as described next. This classification will be used in the subsequent analyses.

A Classification of 12 Environmental Organizations

Detailed and individual analyses of the 12 voluntary organizations permit comparisons beyond the scope of this book. In order to facilitate analysis and to improve our understanding of the environmental field, we develop a fourfold typology, or a classification scheme, into which we classify the 12 organizations. The scheme consists of two dichotomies. The first dichotomy separates "old" from "new" organizations using the year 1985 as the cut point. This criterion rests on the structural change that took place from the mid-1980s most prominently in the environmental field, in which a large membership base and internal democracy became less important. Many organizations formed after 1985 are market-oriented and/or direct-action organizations. Some act as environmental consultants for various businesses. They willingly use the mass media to exert influence on public opinion and government policies. They keep a professional staff and take minor interest in recruiting ordinary members. The absence of local branches and traditional membership classifies these organizations as being not democratic. They are part of an important new direction of the voluntary sector at large (Wollebæk & Selle, 2002a).

The second dichotomy separates core organizations from noncore organizations. Coreness is defined by the following criteria: The organization is of a certain size and influence, it employs a broad definition of environmentalism that, in turn, is its prime task, and it operates in more than one area. This classification is also based on interviews with leaders of the 12 organizations (Strømsnes, 2001). The leaders' assessments of coreness were surprisingly consistent across organizations. Our classification of coreness does not exclude any organization that defines itself as belonging to the core, except NOAH (see discussion below).

The organizations are classified in Table 3.1. The classification is accompanied by the organization's acronym and year of foundation, as well as issue orientation, organizational structure, and work methods. We consider the classification of organizations being old or new as uncontroversial. The coreness dichotomy requires more elaboration. We approach it by classifying the noncore organizations, of which five are identified.

The Norwegian Mountain Touring Association's prime concern is to conserve nature and facilitate outdoor activities. Many sign up as members because they obtain a discount when they use one of the organization's numerous mountain cabins. Although conservation of nature has received increased interest in the organization in recent years, it falls short of being classified as a core organization. The Norwegian Organization for Ecological Agriculture and NOAH—for animal rights are not classified as core because they are organizations of single

TABLE 3.1. Environmental organizations by coreness and age.

	Core	Noncore
Old	**The Norwegian Society for the Conservation of Nature** (NNV: 1914/1963) Broad environmental concern Democratic Cooperates closely with environmental authorities	**The Norwegian Mountain Touring Association** (DNT: 1868) Outdoor activities, mountain-hiking Democratic
	Nature and Youth (NU: 1967) A sensible use of nature Democratic, action-oriented	**The Norwegian Organization for Ecological Agriculture** (NØLL: 1971–2001) Ecological (organic) agriculture
	World Wide Fund for Nature (WWF: 1970) Wildlife, sustainable development International, oriented towards science and lobbying Nondemocratic (except between 2000 and 2005)	Democratic NØLL continued as part of Oikos as of 2001
	The Future in Our Hands (FIVH: 1974) Third world solidarity, global issues, quality of life Democratic	
New	**The Bellona Foundation** (MB: 1986) Issue-oriented, pollution Nondemocratic, action-oriented "Technophiles"	**NOAH—for animal rights** (NOAH: 1989) Animal rights Action-oriented Nondemocratic
	Greenpeace Norway (GN: 1988–1998) Global issues, resources, issue-oriented. International, nondemocratic, action-oriented Merged with its Scandinavian branches 1999/1998	**Women–Environment–Development** (KM: 1991–1997) Female environmentalism Flat organizational structure Folded in 1997
	Green Warriors of Norway (NMF: 1993) Everyday environmentalism and environmental "war" Only activists can vote Action-oriented, nondemocratic	**The Environmental Home Guard** (MHV: 1991) Consumer environmentalism Participants (including organizations) A voluntary national working bee

Note: The distinction between old and new organizations will to some extent coalesce with a distinction between democratic and nondemocratic organizations, with the exception of World Wide Fund for Nature, which was both "old" and nondemocratic when the survey was executed. The democratic member organization was established in 2000, but dissolved in 2005.

issues that carry a too narrow definition of environmentalism. The failure to include NOAH—for animal rights as a core organization also stems from animal rights not being an issue in the broad but moderate Norwegian brand of environmentalism. During interviews, representatives from all 11 organizations other than NOAH failed to perceive it as a core organization. Some even failed to see it as an environmental organization at all (see also Chapters 8 and 9), which implies that the local community perspective not only is strong within the Norwegian population but also within the environmental movement. To NOAH, this was a sore point because it wanted its organization to be perceived both as environmental and as core. The sheer size of Women–Environment–Development prevents it from being a core organization, and the exclusion from coreness is underscored by it entering a sleeping stage in 1996 and being disbanded in 1997 (Gulichsen, 1996). The Environmental Home Guard is an umbrella organization that is not organized as an ordinary voluntary organization. In addition, it does not organize its participants beyond consumer environmentalism and is, therefore, not included as a core organization.[62]

Following the above criteria of coreness, none of the remaining seven organizations can be excluded from the core category. However, the classification of four of these organizations as core might not be self-evident. The Future in Our Hands extends beyond being an environmental organization in that it also engages in developmental issues and international solidarity. These issues are, however, strongly integrated in the overall ideology and rhetoric of the organization. Despite it being a relatively new organization, subsequent organizational development leaves little doubt that Green Warriors of Norway also is a core organization. It is involved in a broad area of environmentalism, where it also influences politics. World Wide Fund for Nature and Greenpeace Norway are national branches of international organizations, primarily through which they gain importance in the Norwegian context. The remaining organizations, The Bellona Foundation, The Norwegian Society for the Conservation of Nature, and Nature and Youth are the three most central environmental organizations in Norway and, without doubt, core organizations.

The four organizations classified as old core dominated the environmental arena for several years when environmentalism, as a policy field, was established in the early 1970s. With the exception of WWF, which had a formal democracy between March 2000 and January 2005 only, all organizations have been democratically built and emphasize the importance of having members.

The two old noncore organizations are both oriented toward a sustainable use of nature (i.e., outdoor activities and agriculture). Both organizations emphasize

---

[62] However, consumerism as political action is, in general, becoming more important (see, for instance, Daunton & Hilton, 2001; Micheletti, 2003; Micheletti, Follesdal, & Stolle, 2004).

that use of nature must not violate nature's own needs. The organizations are member-based and democratically structured.

New core organizations were founded when environmentalism gathered momentum during the late 1980s and early 1990s. These organizations use direct actions to address and solve issues. The role of members has diminished and the leadership is eager to attract media attention. These organizations are less hostile to market mechanisms and market actors and have started to question the environmental contributions of the government in Norwegian society. We are dealing with new core organizations in this important policy field that break with the traditional way of organizing. These organizations advocate attitudes that are not always in congruence with what our anomalies would predict.

New noncore organizations were founded in the late 1980s and early 1990s also. They have an organizational structure similar to that of the new core organizations, only expressing the more general transformation of the voluntary sector. Each of the three organizations promote a single issue (i.e., animal rights, female environmentalism, and consumer environmentalism). In a sense, they represent the increased specialization that is also found in the voluntary sector at large (Wollebæk & Selle, 2002a). Thus, these organizations offer new and separate issues to Norwegian environmentalism.

The typology is also based on assessments of type of environmental coherence at the organizational level, and it offers a parsimonious approach to the analysis of the environmental movement. The typology offers criteria to reduce variance and it brings up front what typifies groups of organizations. One drawback of this approach is that one might restrain variance and fail to observe important variation between organizations within a type. Our analytic procedure, therefore, is, first, to consider the main differences between the general population and organized environmentalists. Second, we consider variations across the four types. Third, we relax the typology and analyze individual organizations as deemed necessary. Also, when necessary for the analysis, we distinguish between democratic and nondemocratic organizations instead of new versus old organizations.

Knowledge about the two anomalies and knowledge about the history and development of environmental organizations are not sufficient to inform us of the nature and character of the types of environmental organization. Next, we approach the environmental organizations by analyzing their members in terms of social characteristics, attitudes, and behaviors. One can never fully understand an organization through knowledge about its members only. However, without such knowledge, one also fails to understand an organization fully. Membership information gives important information about what kind of organization we are talking about and the kind of space within which the leadership operates. Of particular interest is to view the extent of "symmetry" between organizational goals and structure, on the one side, and members' attitudes and behaviors, on the other. That is what we will turn to in Part II of this book.

Part II:
The Environmentalists.
Who They Are, How They Think, and How They Behave

Chapter 4
Who Are the Environmentalists?

Introduction

An organization consists of members who, in the aggregate, contribute to the characteristics of the very organization. The overall character of an organization is not arbitrarily linked to whether its members consists of one gender only, whether age distribution is highly skewed compared to the general adult population, or whether members are exclusively drawn from top income brackets. Research on individuals who are environmentally concerned and on members of environmental organizations conclude that they, in general, are younger, better educated, and more radical than the general population (Lowe & Rüdig, 1986; Van Liere & Dunlap, 1980). Although the age factor seems to have become less salient over time due to possible cohort effects (Mertig & Dunlap, 2001), the general assumption in environmental studies is still that members of environmental organizations differ from the general population. However, within our Norwegian case of unique environmentalism, the two anomalies—the local community perspective and state-friendliness—lead us to expect that members of environmental organizations are not too different from people in general. Where appropriate, we also draw on other publications on the voluntary sector in general because this body of work leads us to expect that organized environmentalists are very similar to members in the voluntary sector at large. Starting with the present chapter and continuing with the two subsequent ones, we compare organized environmentalists with the general population with respect to their sociodemographic characteristics, attitudes and beliefs, as well as environmental and political behaviors.

Gender

The relationship between gender and environmental involvement is ambiguous. According to an ecofeminist position, women and nature are *positively* united by means of their reproductive capacity. Women who become mothers have an embodied advantage that is unavailable to men. This is referred to as the "body-based argument." An ecofeminist position also refers to an "oppression argument," which states that women and nature unite negatively in their subordinated positions in the great patriarchal hierarchy: God–man–woman–nature (Eckersley, 1992; Warren, 1994). These two arguments lead us to expect that women should be more likely than men to join environmental organizations. Often, it is also claimed that environmentalism is a "soft" policy area and thus appeals more to women than to men (Skjeie, 1992; Togeby, 1984). Contrary to this, researchers have found that men more than women are likely to be politically active and more involved in community issues. Hence, one might expect that men also will participate more in environmental organizations. On the other hand, men often hold jobs that depend more directly on economic growth and technological sophistication. Because environmental organizations often are understood as being part of the new social movements that often align themselves against this type of modernization, men are less likely to support issues that relate to the environment (Dietz, Kalof, & Stern, 2002; Hunter, Hatch, & Johnson, 2004; Mohai, 1992; Tindall, Davies, & Mauboules, 2003).

Our analysis shows that women are slightly more likely to be a member of environmental organizations. Among organized environmentalists, the gender distribution is 51% women and 49% men. Were it the case that environmentalism attracts women, or women being drawn toward environmentalism, we would have expected to find a stronger difference between the genders.[63] When looking at different types of organization, the data show that women are more likely to be a member of new environmental organizations, whereas men are more often found in old environmental organizations. On the other hand, men are more likely to join core organizations, whereas women are more likely to be a member of a noncore organization.

Framed within our typology of unique environmentalism in Table 3.1, men are most often found in environmental organizations in the positive diagonal (i.e., new core organizations and old noncore organizations), whereas women predominate in new noncore organizations.[64] There are no gender differences in old core organizations. Specifically, women are found more frequently in the new noncore organizations The Environmental Home Guard and NOAH—for animal rights (63% and 75%, respectively), most likely because of the consumer environmentalism of the former (women still have the main responsibility for domestic work and housekeeping) and the protection of the weak (animals) of the latter (see, e.g., Jasper & Nelkin, 1992).

---

[63] The organization Women–Environment–Development was removed from this analysis because it recruits women only; when this organization is included, the gender difference increases. We do not know the true distribution of gender in the other 11 organizations due to lack of population data. In the general population sample, the gender distribution is 48% women and 52% men, as more men than women returned the questionnaire.

[64] Women–Environment–Development was removed from this analysis also.

Age

We assume that environmental organizations are somewhat radical but not necessarily subversive, aiming to be a challenger and corrector to the state. We also acknowledge the kernel of truth in the adage that the hearts make young people radical, whereas the brains and wallets make old people conservative. Basically, young people are expected to be "less integrated into the dominant social order and therefore more ready to accept solutions to problems which may require substantial changes in traditional values and behavior" (Lowe & Rüdig, 1986). Members of environmental organizations should be more likely to be considerably younger than the populational average.[65] The validity of the assumption should increase when addressing organizations that advocate unconventional political behavior and endorse direct action. However, again, such differences should be tempered in a setting in which the two anomalies operate.

The data show that the average age of members of environmental organizations is 41 years, whereas the average age in the general population sample is 44 years.[66] Although members of such organizations are younger than the general population, we are struck by the insignificant age difference of 3 years. This result indicates that organized environmentalists in Norway are surprisingly old, given the hypothesis that members of such organizations should be "less integrated into the dominant social order." This can also indicate that the Norwegian environmental movement is not much of an alternative movement. On the other hand, the result corresponds with the general observation that people of middle age are more likely to participate in politics than younger and older people (Strømsnes, 2003; Wollebæk et al., 2000).

We find the oldest members in old noncore organizations as well as in new core organizations (the average age is 43 years for both types of organization). This result is interesting, given that the activist approach found in new core organizations should appeal more to younger environmentalists. However, the results not only indicate that new core organizations recruit rather mature and seasoned environmentalists, but the results might also be explained by the fact that members of these organizations are passive members and in a sense pay their dues so that others can have the means to act. The youngest environmentalists are found in old core organizations (the average being 39 years) as well as in new noncore organizations (the average is 40 years). However, if we exclude Nature and Youth and NOAH—for animal rights from the analysis (both organizations attract young members), no significant age differences are found between the types of organization or between the organized and the general population.

---

[65] We hasten to add that the overall picture is somewhat more complex than what we portray here as generation effects might interact with life-cycle effects. Contemporary youth seems to be less radical than their preceding cohorts, at least compared to those that were socialized back in the 1960s (Wollebæk & Selle, 2003).

[66] The standard deviations are approximately 16 and 18, respectively.

Norwegian environmentalists today are rather mature individuals. The failure to observe significant age differences between organized environmentalists and the general population suggests that environmental organizations have failed to attract and keep in their ranks younger people. This might stem from both that young people do not possess enough radicalism to be lured by environmental organizations and that the organizations are unable to present themselves as radical alternatives to the state. The state-friendly society thesis might very well account for this indistinction between organized environmentalists and the general population.

Education

If education generates political interest and participation, the field of environmentalism cannot be left untouched by the effect of this kind of cultural capital. Thus, people with higher education are more likely to be concerned about the environment and to join environmental organizations. The environmental field in part draws on scientific traditions that appeal to those with higher education. People with higher education are, to a greater degree, exposed to environmental information, and they have a greater ability to acquire and understand environmental complexity. Higher education "helps to cultivate the ability to think critically, question everyday assumptions [and] form an independent judgement" (Eckersley, 1989, p. 221). People with lower education tend to be less able to sort out the, at times, overwhelming and often contradictory details of environmental information and rationalize lack of involvement by pointing to the miniscule effect of one individual's action.

Education is, along with employment and income, an important factor in making up the "new class," one of the virtues being its "relative autonomy from the production process" (Eckersley, 1989, p. 221; 1992, p. 63). Lowe and Rüdig (1986) also formulated a "relative deprivation" hypothesis in which they stated that "the better educated experience superior environmental conditions at work and in their leisure time and are therefore likely to be more sensitive to environmental deterioration." It is therefore reasonable to expect, given this tradition and without controlling for other variables such as exposure, that higher education reduces the possibility for environmental neglect (Freudenburg & Gramling, 1989).[67]

We measure education as number of years beyond compulsory school.[68] Because education can be considered as a type of "cultural capital" that, to some degree, can be desired by one's parents, we also include the mother's and father's

---

[67] The effect of education can be modified under the influence of other variables. See our discussion later in this chapter as well as in Chapter 7.

[68] In Norway, compulsory education was 7 years until the middle of the 1960s, 9 years until 1996, and 10 years since 1997.

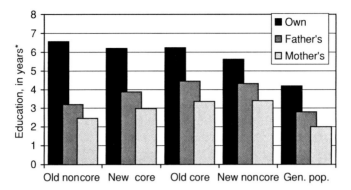

FIGURE 4.1. Education and organization type.
*Education is measured in years beyond compulsory school.
N_{GP} = 872/589/592, N_{OE} = 1889/1474/1473.

educational level. In Figure 4.1, a value of 3 is equivalent to having completed the gymnasium (or high school), whereas a value of 7 is equivalent to having completed the lowest university education (a bachelor's degree).

Organized environmentalists' own education is approximately 2 years longer than that of the general population. This result confirms our overall assumption. The result translates roughly into the average education of the population being a first-year college education, whereas the average education of members of environmental organizations is almost a completed lower university degree. Figure 4.1 shows that members of old noncore organizations have an approximately 1-year longer education than members of new noncore organizations. Inversely, the parents of members of new noncore organizations have more education than parents of members of old noncore organizations. The educational level of parents of organized environmentalists outperform that of the general population: both the mother's and father's education are approximately 1 year longer.[69]

These results confirm that education and environmentalism go together, as do education and voluntary participation in general (Wollebæk et al., 2000). There exists not only an undeniable element of cultural capital among members in environmental organizations, but parental cultural capital is also strongly present. Although several of the aspects linked to the above hypotheses are not directly tested here, our results do not deny that education is related to environmental concern and participation when measured as membership in environmental organizations. If follows that higher education reduces the possibility for environmental neglect.

---

[69] We also analyzed the relationship between education and organizational types when controlling for age. Organized environmentalists have longer education than the general population in all age categories. From the age of 25 and up, the difference in education between the general population and organized environmentalists increases.

Income

Education, occupation, and income are intimately related as indicators of the new middle class. Income alone is an indicator of the affluence of the new middle class. The new middle class is the locus for the theory of postmaterial value shift. This theory holds that when material well-being is sustained during one's formative years, new generations can give priority to postmaterial values such as equality, political participation, and a clean environment (Inglehart, 1977, 1997). Higher income—also referred to as economic capital—enables individuals to spend proportionately less on material necessities such as food and shelter.

Measured as the proportion of respondents having personal and household incomes exceeding NOK 300,000 (approximately USD 42,000 in 2003), analysis shows distinct income difference between organized environmentalists and the general population, as well as within the environmental movement. Twelve percent of organized environmentalists, as opposed to 10% of the general population, have a personal income exceeding NOK 300,000. The figures for household income are 54% and 48%, respectively.

Members of old organizations have greater personal income than do members of new organizations, whereas members in core organizations earn more than members of noncore organizations. Members of new noncore organizations are least frequently found with high personal incomes; only 3% of these members have a personal income above NOK 300,000. Sixty percent of members of old core organizations have a household income above NOK 300,000. The proportions for the members of organizations in the other three types vary between 48% and 52%. Members of World Wide Fund for Nature are most frequently found in the top income brackets, in that 72% of its members report a household income above NOK 300.000.[70]

The bottom line is that income, added to the previous indicator of education, shows that the environmental movement can be safely located within the middle class. Members of environmental organizations not only exceed the general population in cultural capital but also in economic capital.

Public Versus Private Sector

Not only might age, education, and income affect one's environmental involvement, employment in the private sector is regarded as a negative and weak indicator of the new social class because researchers have shown a positive

---

[70] When controlling household income for age, organized environmentalists increase the income gap among respondents aged 50 and higher compared to the general population.

association between employment in the public sector and environmental concern and behavior (Martell, 1994; Van Liere & Dunlap, 1980). The analyses show that 55% of the general population and 43% of organized environmentalists are employed in the private sector. Although a strict definition of the public–private distinction might be difficult to obtain, the present definition is, of course, constant across the analytical categories.[71] The results are therefore conspicuous. The public–private distinction is striking even between the types of environmental organization. Private employment is found among 49% of members in new core organizations, 47% for old noncore organizations, 41% for old core organizations, and 34% for new noncore organizations.

The expected divide of sector employment might be explained by the claim that environmentalists to a larger extent than the general population opt for jobs in the public sector or in organizations because they are less attracted to employment directly related to economic markets and industries. Public-sector employment is conducive to a general environmental cause because, after all, it is the public sector that implements limitations on free markets and imposes restrictions on businesses and industries. The mechanisms and directions of cause and effect might be difficult to disentangle. We interpret the magnitude of numbers on sector employment to support our state-friendliness hypothesis. Were the state understood as a reluctant and intransigent actor, the differences in occupational numbers between environmentalists and the general population would have been smaller.

Residence

The local community perspective of Norwegian environmentalism might interfere with the general urban–rural dichotomy found in the literature on environmentalism. This perspective, emphasizing the development of vital and self-sufficient districts and local communities, is at the center of Norwegian politics. Norwegian environmentalism is heavily influenced by this perspective because it emphasizes the survival of local communities. This means that nature, as a means of sustenance, is given precedence over the conservation of nature for nature's own sake.

As city dwellers, urban residents are more concerned with pollution because they simply are more exposed to it. Urban residents are also expected to be less involved in extractive occupations. Rural residents, on the other hand, would, in general, be less exposed to pollution because to a lesser degree it is found in the countryside. However, rural residents might be involved in, or closer to, extractive occupations, from which a more utilitarian approach to nature might emerge. Urban and rural residents might be very crude categories, but they more often

---

[71] The nonprivate sector includes employment in municipal, county, or state bodies, as well as government-related but more independent organizations and institutions.

than not coincide with nonextractive versus extractive occupations. In addition, urban residents might correlate with a high perceptability of environmental problems, whereas rural residents might correlate with a low perceptability of environmental problems (Berenguer, Corraliza, & Martín, 2005; Freudenburg, 1991; Van Liere & Dunlap, 1980).

Here, we measure the rural–urban distinction by way of a 5-point scale running from "large city" (1) through "suburban area" (2), "town" (3), "village" (4) to "a less densely populated area" (5). Analyses show that members of environmental organizations receive an average score of 2.8, whereas the general population receives a score of 3.3. This means that environmentalists as a whole are more urbanized than the general population. Whereas the average Norwegian lives somewhere between "town" and "village" and somewhat closer to the former, the average environmentalist lives in the outskirts of suburbia. When we control for our four organizational types, the organizational type that is least urbanized is old noncore (3.3), followed by old core and new noncore organizations (2.8). Members of new core organizations are most urbanized (2.6). New core organizations are also the type of organization that most willingly advocates direct-action environmental behavior.

The respondents were also asked about the degree to which they perceive environmental problems in their neighborhood (i.e., air pollution, traffic noise, poor water quality, garbage, and dying forest). The five issues were transformed into a single scale running from no problems (0) to a perception of all problems (1). On this pollution-perception scale, the general population has a mean of 0.35 and organized environmentalists have a mean of 0.40. It is not surprising that organized environmentalists observe more environmental problems in their neighborhood than do the general population, as environmentalists are more concerned about the environment in the first place. One of the consequences of joining (and reasons for joining) an environmental organization is a diminished neglect of environmental problems. Within the organizational typology, members of new organizations observe more problems than do members of old organizations. The scores are 0.42 and 0.38, respectively. There is no difference between core and noncore organizations.

When perception of environmental problems is controlled for residence, the results show that perception of environmental problems increases with degree of urbanity for both environmentalists and the general population. In rural areas, environmental perception for the general population and organized environmentalists are 0.28 and 0.31, respectively. The scores for cities are 0.40 and 0.47, respectively. It is not surprising that urban dwellers perceive more environmental problems than do rural dwellers because some of the scale items, such as air pollution and traffic noise, are less predominant in rural areas. However, the scores suggest a weak interaction effect between urbanity and environmental perception within the two groups. Moving from rural areas to cities, environmentalists' perception of environmental problems increase more rapidly than do nonenvironmentalists.

Conclusion

Environmentalists have been characterized by Milbrath (1984) as the "vanguard for a new society." In this chapter, we have seen that such a Norwegian vanguard, in terms of sociodemographic characteristics, deviates less from the general population than what environmental literature might lead us to expect. Organized environmentalists do not deviate substantially from the general population. Apart from sex and age, our expectations are confirmed. True, features of the new middle class—to which environmentalists are closely linked—were conspicuously present among environmentalists. On both education and income, environmentalists surpass the general population. This distribution of cultural, social, and economic capital is also found when the general population is compared with the whole voluntary sector (Wollebæk et al., 2000). On the tried and tested, but perhaps somewhat attenuated indicator of age, the organized environmentalists failed to emerge as the youthful force that one should expect of an ostensible alternative social movement. This result seems to coincide with the observation that environmental involvement drew heavily on those who were young in the 1960s and 1970s and without being able to recruit subsequent cohorts. The results also seem to corroborate the expectations of the lack of differences that will be observed between the general population and organized environmentalists when the two anomalies are in effect.

Chapter 5
Environmentalists Without an Attitude

Introduction

When a new issue dimension emanates in the political field, scholars make attempts to identify its nature and structure: What is the content and how is it similar to or different from what we already know? In this vein, some scholars consider environmentalism as a more or less coherent ideology (Dobson, 1993), whereas others claim that its structure might consist of more than one dimension (Oelschlaeger, 1991; Paehlke, 1989; Young, 1992). A key question that always remains is to determine how the new cluster of issues relates to other conventional and existing political dimensions. For instance, when drawing on respondents' wide array of environmental beliefs and attitudes, one can determine whether environmentalism is independent of the conventional radical–conservative distinction as in the phrase "Neither Left nor Right, we are out in front" (Dalton, 1994, p. 122) or the degree to which environmentalism overlaps with the conventional distinction.

In this chapter, we compare the beliefs and attitudes of organized environmentalists with those of the general population, using four frameworks.[72] The first framework consists of the conventional left–right distinction and preference for political parties. To this we also add questions on the relevance of the left–right distinction, the EU issue, and the ostensible perennial dilemma of economic growth versus environmental protection. Second, we use Inglehart's (1990, 1997) postmaterialist framework of value change to see whether this set of values can be discerned among members of environmental groups. Third, we use grid–group cultural theory to determine whether cultural values are unequally distributed within and between the environmental members and the general population (Ellis & Thompson, 1997a; Thompson,

---

[72] We use the term *framework* because they differ as to dimensionality and theoretical complexity.

Ellis, & Wildavsky, 1990; Thompson, Grendstad, & Selle, 1999). Fourth, we employ both the New Environmental Paradigm Scale developed by Dunlap and associates (Dunlap & Van Liere, 1978; Dunlap, Van Liere, Mertig, & Jones, 2000) and Eckersley's (1992) framework that ambitiously attempts to identify more subtle distinctions between anthropocentric and ecocentric beliefs. To these distinct frameworks, we also subjoin measurement of technology and religion because of their ideological components and their interesting links to the other measures.

We use these four frameworks somewhat eclectically in order to improve our understanding of the organized environmentalists and under what conditions organized environmentalism operate. With due respect to the dictum that nothing is more practical than a good theory, we believe that a more rigorous use of, and commitment to, the present frameworks would obstruct the subtleties of our anomalous case of organized environmentalism to emerge from the data. According to the state-friendly society and local community hypotheses, we generally expect the core organizations to be found toward the moderate positions of these frameworks. This stems from our belief that the close and existing cooperation between the core groups and the state would never materialize if environmentalists were politically and ideologically strongly polarized and in deep conflict with government bodies. In turn, the existing cooperation also tempers the political positions of the core groups.

From Right to Left

Many aspects of environmentalism are reached by the tentacles of what for a long time has been understood as conventional politics. It is understandable that new issues are attempted to be understood by way of conventional thinking. Some aspects of environmentalism do not sit easily with conventional politics. Some positions have even changed over the years.

Historically, only a relative material independence from the often harsh and uncompromising demands of life could ensure a foundation from which interest in nature in its own right could develop. The preservation of nature and conservation of resources were therefore often linked to conservatives and the contemporary political right. Today, concerns for nature and a commitment to environmentalism cross-cuts political positions. A quick solution to environmental problems requires an active government (Dunlap, 1995; Lowe & Rüdig, 1986), whereas environmental policies often implies a rejection of the status quo (Dunlap & Van Liere, 1978). Because business and industry generally oppose environmental reforms due to the costs involved and because environmental reforms entail extending government activities and regulations, it is more likely to find environmentalists among the radical left than among the conservative right. Previous research has shown a leftist leaning among organized environmentalists (Dunlap, 1975; Ellis & Thompson, 1997b; Jones & Dunlap, 1992). If we locate the environmentalists further to the left than the general population, the independence between the left–right radicalism and environmentalism must be questioned.[73]

The survey included a 10-point left–right self-placement scale. A score of 1 indicates the extreme left and a score of 10 indicates the extreme right. The midpoint of such a scale is 5.5. We found that the general population is located squarely in the political center (5.4), whereas organized environmentalists are found to be significantly more to the left (4.3).[74] This result supports previous research that environmentalists are more leftist than the general population. The result also leads us to anticipate a rejection of the expectation that environmentalism and the left–right dimension are independent of one another.

Are there reasons to assume that organized environmentalists reject the validity of a left–right distinction? We do not think so. The fact that a large proportion of the respondents were willing to place themselves on this scale indicates that it has a high validity. A very high degree of "recognition" of the scale (Fuchs & Klingemann, 1989; Klingemann, 1995) is evident because 95% of the environmentalists and 91% of the general population placed themselves on the scale. We explicitly asked the organized environmentalists to evaluate whether the left–right scale represent the political landscape well or poorly.[75] Roughly half of the members of environmental organizations, regardless of organizational type, agree with the statement that the left–right dimension describes the political landscape poorly. However, our analyses show that rejection of the left–right scale does not influence left–right positioning among the members of different organizational types.

The left–right dimension is often used to delimit the political space of a political system. In representative democracies, this political space is occupied by political parties. A striking feature of Norwegian politics is the absence of a competitive green party in what is considered a general green polity.[76] In electoral politics, the Norwegian green party never experienced any success (Grendstad & Ness, 2006; see also Chapter 8). It is therefore more often than not excluded from the list of the roughly eight parties of electable size when the general population's party preferences are surveyed. Therefore, the emerging question is: What is the party of choice for the environmentalists?

Compared to the general population, organized environmentalists prefer political parties found at the political left more than parties found in the political center or at the political right. The Socialist Left Party is most popular among the environmentalists and attracts three times as many environmentalists

---

[73] On the contestedness of the left–right dimension, see Kitschelt and Hellemans (1990), Knutsen (1995), and Grendstad (2003a). The studies hold that the left–right dimension shows great flexibility in incorporating new issues and adapting to new policies.

[74] Differences of the means among the four types were not statistically significant.

[75] The question was not asked to the general population. See also Chapter 8.

[76] For the various ratings and rankings of environmental performances and greenness, see Esty and colleagues (2002, 2005).

than voters from the general population (see Figure 5.1).[77] Knowing that the Socialist People's Party was founded in 1960 and for decades was the foremost representative for the Norwegian variant of left-wing populism, we can easily relate preferences for the Socialist Left Party to the local community perspective.

The Liberal Party and the Red Election Alliance are also more popular among environmentalists than among the general population. The general population supports the Liberal Party with 2.2%, whereas the environmentalists support the Liberal Party with 12.2%. This gives a factor of 5.5. The factor for the Red Election Alliance is 4.7. The Labour Party is underrepresented among organized environmentalists by a factor of 0.5 (i.e., 15.9% for environmentalists divided by 34.2% for the general population). Underrepresentation is also present for the Progress Party (0.4), Christian People's Party (0.5), the Conservative Party (0.7) and the Center Party (0.8), all of which are considered center parties or right-wing parties.[78] By preferring radical political parties as well as taking a more radical position on the left–right scale, we can safely conclude that organized environmentalists are found more to the politically left than the general population. However, we still believe that the differences

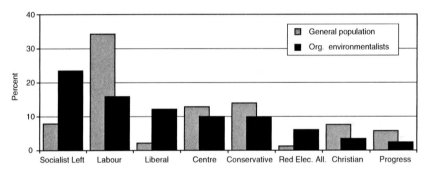

FIGURE 5.1. Party preference.
$N_{GP} = 959$, $N_{OE} = 1942$.

---

[77] The "other party" and "don't know" categories are not included here. For the general population, these figures are 1% and 13%, respectively, and for organized environmentalists, they are 4% and 13%, respectively. We are unable to determine the extent to which the 4% includes votes for the Norwegian Green Party (see also Chapter 8). There are insignificant variations of party preference across organizational types.

[78] The leftist leaning among organized environmentalists is not found for members of voluntary organizations in general. Wollebæk and colleagues (2000, p. 194), for instance, found that those who are most positive to the work performed by voluntary organizations are overrepresented in the political center (i.e., in the Christian People's Party, the Liberal Party, and the Center Party). Those most in favor of market solutions are, not surprisingly, found among the rightist parties, whereas approval of state solutions and a leftist leaning tend to coalesce. Thus, organized environmentalists are untypical for the voluntary society in their political leanings. Interestingly, the Liberal Party is overrepresented both among voluntarists in general and among environmentalists. This illustrates the strong position that voluntary work has among liberal voters.

between environmentalists and the general population on ideological position and party preferences should have been much larger were environmental organizations to be understood not only as an alternative but also as a strong challenger to the state and the society.

The failure of the Labour Party, the Conservative Party, and the Progress Party to attract environmental voters is basically that they come across as parties of the establishment: They give priority to economic growth over environmental protection. The survey item tapping the positions of economic growth versus environmental protection asked simply if one position should be given priority over the other, or whether the two positions could be combined. The results show that two-thirds of the general population agree that growth and protection can be combined, whereas one-third of all organized environmentalists also hold this position. On a similar question posed to organized environmentalists only, but to which no option of compromise between economic growth and environmental protection was offered, 96% rejected the claim that employment and economic competition should be given priority over a cleaner environment.[79] Thus, it is only in absence of a compromise option that environmentalists' priority of the protection of the environment completely overrides the notion of economic growth. We find that the environmentalists' easy slide into a compromise option is a general sign of moderation on the part of the environmental movement.

In the same way that pollution is a cross-boundary issue, so are the environmental concerns held by the organized environmentalists. However, internationalization poses a general dilemma for environmentalists. On the one hand, increased cooperation and trade might also facilitate the context in which international bodies and actors might seek to solve the environmental problems that more often than not defy national borders. On the other hand, it might also make it easier for the very businesses and organizations that might be responsible for many environmental problems to move from one country to another where legislation might still be lax. The local community perspective will make Norway's environmentalists anxious about too great an international commitment. Specifically, the part of internationalization that is referred to as Europeanization, or European integration, has always been contentious in Norway (Selle & Østerud, 2006).

European integration has haunted Norwegian politics since the 1960s, during which period the country has, in practice, applied for full membership four times (1961–1962, 1967, 1970–1972, and 1992–1994), reaching the point of finalizing accession terms twice, and rejecting full membership in a public referendum in 1972 and 1994 (Miles, 1996). In the consultative Norwegian November 1994 referendum on membership in the European Union, 47.7% voted "yes" and 52.3% voted "no." On a question in our survey on what the respondent voted in that

---

[79] There were no significant differences among different organizational types on these two questions.

referendum, 53% of the general public in the sample reported to have voted "no," whereas 67% of organized environmentalists reported to have voted "no."[80] The large majority of "no" voters among members of environmental organizations might be at odds with EU-friendly environmental movements in other European countries (Rootes, 2003). In part, we think that this illustrates a weak international orientation within the Norwegian environmental movement that can be considered a consequence of the local community perspective: Norwegian organized environmentalists are within a nation-state setting primarily oriented towards the local communities.[81]

Taken together, our analyses show that political differences to a greater extent run between the general population and organized environmentalists than between the types of organized environmentalist. However, we believe that the differences between the general population and organized environmentalists nevertheless are small due to the effects of the local community perspective and the state-friendly society. Even though the environmentalists place themselves further to the left on the left–right scale, environmentalitsts still recognize the scale. Even though the environmentalists vote more frequently for parties on the traditional left—but this does not include the Labour Party—they do not vote that differently from the general population. Even though they favor environmental protection over economic growth to a greater extent than does the general population, the environmentalists too easily slide into the compromise position once it is made available. These positions prevent the organized environmentalists from presenting themselves as an alternative to state and society. They are not on the outside of ordinary party politics. Even less do these positions enable organized environmentalists in any way to constitute a national societal subculture of their own.

Postmaterialism

The theory of postmaterial value change (Inglehart, 1971, 1977, 1979, 1981, 1990) is a familiar, albeit contested, approach to the concepts of new politics and the general value change that has gained importance in Western industrialized countries after World War II (Clarke, Kornberg, McIntyre, Bauer-Kaase, & Kaase, 1999; Davis & Davenport, 1999; Flanagan, 1987; Grendstad & Selle, 1999; Inglehart & Abramson, 1999). Inglehart argues that the change reflects the postwar generation's move away from materialist values such as political order and economic stability and toward postmaterialist values such as political participation and increased influence on government decisions.

---

[80] There are no significant differences among types of environmental organization.

[81] The Rootes study (2003) showed that environmental protest is still very national in character within the EU countries. There is also a lack of coordination across countries. However, we are talking about something more than protest: the cognitive or ideological orientation toward what is national or international (see Chapter 2).

The materialist–postmaterialist theory rests on two conjoined hypotheses. First, the scarcity hypothesis states that individuals tend to place high priority on whatever is in short supply. Second, the socialization hypothesis claims that individuals tend to retain a given set of value priorities throughout adult life once this set has been established in their formative years. Initially, to these hypotheses Inglehart added Maslow's (1954) hierarchy of needs, which states that physiological needs precede higher-order needs, in order to provide his theory with the direction of value change. Thus, the formative experience of economic and political security makes it highly probable that an individual will keep postmaterialist values in adult life. Later, Inglehart replaced Maslow's hierarchy of needs with a concept of diminishing marginal utility (see Grendstad & Selle, 1999). This replacement permitted the theory to work both ways in that the one-way street from materialism to postmaterialism was turned into a two-way street between the two conceptual positions. This change introduced greater theoretical flexibility (Inglehart, 1997).

Originally, environmentalism was nothing more than a passing footnote in the theory of postmaterial value change (Inglehart, 1971, p. 1012). Later, green politics was considered "the archetypical example of postmaterial politics" (Dalton, 1994, p. xiii; Inglehart, 1990, p. 267). One must, therefore, presuppose that postmaterialism and being an environmentalist to some degree coalesce, as both of these recent bodies of thought and behavior draw heavily on a value change (Franklin & Rüdig, 1995). Specifically, the postmaterialist theory argues that it, in part, embodies a transformation of the traditional left–right dimension so that the major conflict in politics today takes place between a postmaterial left and a materialist right (Inglehart, 1990).

Inglehart used people's priorities of four policy goals in a two-step process to construct the materialist–postmaterialist index. Postmaterialists favor increased democratic participation and the protection of the freedom of speech, whereas materialists find maintaining law and order and fighting rising prices to be more important. For the present analyses, we have followed standard procedure and made four categories (i.e., strong materialist, weak materialist, weak postmaterialist, and strong postmaterialist). A strong materialist is a person who has a materialist item on the first and second priority. A strong postmaterialist is a person who has a postmaterialist item on the first and second priority. A weak materialist mixes a material and a postmaterial item, but with the materialist item first. A weak postmaterialist does the same but in reversed order.[82]

---

[82] The four policy goals are (1) maintaining law and order in the country, (2) giving the people more of a say in political decision-making, (3) fighting inflation, and (4) protecting freedom of speech. Issues 1 and 3 represent materialism; isues 2 and 4 represent postmaterialism. Because the materialist–postmaterialist items were not posed to the general population in the 1995 Survey of Environmentalism, results for them are taken from the *1993 International Social Survey* in Norway. In the 2000 International Social Survey of the general Norwegian population, the percentages of pure materialists and postmaterialists were identical to those obtained in 1993. We are grateful to the Norwegian Social Science Data Services, University of Bergen, for allowing us to use these data; the responsibility for the interpretation of these data rests solely with the authors.

The materialist–postmaterialist distribution for the five groups shows that 20% of organized environmentalists are strong postmaterialists (see Figure 5.2). This is twice the proportion of strong postmaterialists among the general population. Twenty percent of the general population are strong materialists. Only 5% of the organized environmentalists are found in this category.[83]

These distributions distinguish organized environmentalists from the general population. Primarily, environmentalists are more postmaterialist than the general population, and the general population is more materialist than environmentalists. However, based on Inglehart's theory, one should expect a stronger postmaterialist tendency among organized environmentalists. Even though the strong and weak postmaterialist categories for the environmentalists add to roughly 50%, it would be wrong to conclude that postmaterialism is a dimension around which environmentalists rally.[84] If it is the case that green politics is considered "the archetypical example of postmaterial politics" (Dalton, 1994, p. xiii) and that "Norway is one of the most postmaterial society on earth" (Inglehart, 1997, p. 22), then it is difficult to see how the results can be taken as support of those claims (see Chapters 7 and 8). However, as we will see, there are other frameworks that address these problems.

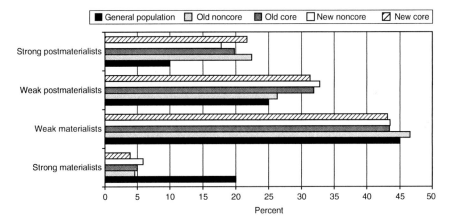

FIGURE 5.2. Postmaterialism.
$N_{GP} = 1414$, $N_{OE} = 1670$.

---

[83] No significant differences are found between organizational types.

[84] These findings would also fit the expectations of the Dryzek study (2003). We do not disagree with their very broad description of characteristics of Norwegian environmentalists. It is their explanation of why Norwegian environmentalists turned out the way they did that we do not subscribe to (see Chapter 2).

Grid–Group Theory

Grid–group theory, sometimes also referred to as grid-group cultural theory or "Cultural Theory," was developed by Mary Douglas, Aaron Wildavsky, and Michael Thompson (Douglas, 1982; Thompson et al., 1990, 1999; Wildavsky, 1987). The name of the theory refers to the dimensions grid and group, which demarcate patterns of *social relations*. Group concerns the degree to which individuals find themselves in a tight-knit group through which feeling of togetherness and solidarity among the members will develop. Grid is the extent of rules and prescriptions that regulate the behavior of individuals. The combination weak group/weak grid is individualism, weak group/strong grid is fatalism, strong group/strong grid is hierarchy, and strong group/ weak grid is egalitarianism. Social relations go together with *cultural biases* and constitute distinct ways of life. Because the cultural biases are rooted in social relations, one can advance them as deep-seated beliefs and political orientations.

We emphasize the typological aspect of grid–group theory and use the four cultural biases: individualism, fatalism, hierarchy, and egalitarianism. The central aspect of cultural biases is found in the biases' varying notions of the concept of equality. These are as follows: egalitarianism: equality of result; individualism: equality of opportunity; hierarchy: procedural equality; and fatalism: "no equality on this earth" (Thompson, 1992). Each notion of equality is used to justify an issue (i.e., position, goal, or policy). These four justifications, grid–group theory argues, are universal, whereas the number of issues available to them is practically unlimited. This distinction enables one to identify how the crucial social mechanism of justification connects the interlinked social relations and worldviews. Although the distinction posits that there will be issues within this mechanism between the worldviews and social relations, specifically which issue that will be (nuclear reactors, windmills, abortion, slavery, vouchers, you name it) is almost impossible to predict. The point is that once an issue is co-opted to justify a way of life, the issue cannot in any way be a constituent of the mechanism. Therefore, we argue that the biases themselves are not easily changed (Grendstad & Selle, 1995).

Grid–group theory lends itself to various research fields (Thompson et al., 1999). For instance, earlier research attempted to explain why individuals choose the risk they do (Douglas, 1972; Douglas & Wildavsky, 1982). Today, the theory seems particularly interesting when applied to environmentalism (Wildavsky, 1995) and it has significantly influenced a state-of-the-art report on human choice and global climate change (Rayner & Malone, 1998a, 1998b).

Our theoretical and empirical expectation is that egalitarian cultural biases are more predominant among organized environmentalists than among the general public. Theoretically, egalitarianism is linked to political participation and social solidarity. Empirically, it is linked to "the new left" politics and new social movements. Environmental activists are worried about environmental problems "not only because they are concerned about the fate of the earth but because they desire to transform how human beings live with one another in an egalitarian direction" (Ellis & Thompson, 1997a, p. 885). We also expect that organized environmentalists will disagree with individualistic and hierarchical biases due to their emphasis on competition between unconstrained actors and inequality imposed by top-down institutions, respectively.

Our survey included two items for each of the four cultural biases (Grendstad & Selle, 2000).[85] The items were added pairwise and composite scores range from 1 (strongly disagree) to 5 (strongly agree). Figure 5.3 shows that organized environmentalists score higher on egalitarianism than do the general public. On every other cultural bias, organized environmentalists score lower than the general population. These results follow the expectations closely, but the results are also striking in two other related ways. First, the results are striking in that the differences between the general population and the environmentalists are small on all biases. Second, specifically we are surprised by the small differences between organized environmentalists and the general population on our index of individualism because individualism with its emphasis on individual pursuit of happiness more than any other culture should be somewhat extreme to the cause of environmentalism.[86] There are small but significant differences between the organizational types across three cultural biases. Only on hierarchy are we able to discern differences. Members of old noncore organizations show strongest deference to authority, whereas members of new core organizations are most willing to challenge authority. The relative action-prone style of new core organizations does not sit well with values of station and subordinance. In addition to grid–group theory not necessarily being the ultimate theory to maximize differences between the organizational groups and the population, we opine that the overall similarities stem from the lack of polarization between the general environmental movement and the population at large. This is linked both to the state-friendly society and the local community theses.

---

[85] *Hierarchy*: One problem with people today is that they too often resist authority; the best we can do for the coming generations is to hold on to our customs and traditions. *Egalitarianism*: We must distribute wealth more evenly so that there is more justice in the world; I am in favor of tax reform that places the largest burden on companies and individuals with a high income. *Individualism*: If a person has enough vision and ability to acquire wealth, then s/he should be allowed to enjoy it; everyone should have an equal opportunity to fail or succeed without the government interfering. *Fatalism*: Only rarely does anything come out of cooperation with others; it seems that no matter what party you vote for, everything carries on as before.

[86] The differences between the general population and organized environmentalists are small but significant ($p < 0.05$) on all biases.

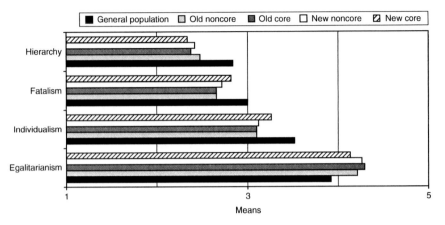

FIGURE 5.3. Cultural biases.
$N_{GP} = 894/950/906/943$, $N_{OE} = 1906/1972/1823/1966$.

Anthropocentrism Versus Ecocentrism

There exist several approaches as to how one can conceptualize and measure ecological attitudes. Here, we will deal with two. First, Dunlap and associates proposed a scale intended to distinguish between the New Environmental, or later
Ecological, Paradigm and the Dominant Social Paradigm (Dunlap & Van Liere,
1978; Dunlap et al., 2000). Second, Eckersley (1992) elaborated an environmental spectrum that polarizes the concepts anthropocentrism and ecocentrism along
which major streams of environmentalism can be discerned. The two approaches
differ as to their theoretical points of departure, but they are similar in their
attempts to map the environmental field.

    The New Environmental Paradigm Scale was developed by Dunlap and asso-
ciates in the late 1970s (Catton & Dunlap, 1980; Dunlap & Van Liere, 1978). In
1992, the scale was reworked and expanded to take account of recent environmental developments. It was renamed the New Ecological Paradigm Scale
(Dunlap et al., 2000). The most recent scale is intended to measure the polarized
worldviews of the Dominant Social Paradigm (DSP) and New Ecological
Paradigm (NEP). DSP rests upon an anthropocentrism in which humans are
viewed as being outside and above nature and in which nature is instrumental for
the benefits of humans. By contrast, NEP contains a rejection of human exemptionalism and continuous economic growth, and a belief that any progress must
pay careful attention to the limitations found in nature itself and the ways in
which humans interact with nature. Previous research claims that "the NEP scale
represents an advanced tool for measuring environmental concern" (Noe & Snow,
1990, p. 26). We assume that organized environmentalists score higher than the
general population on the NEP scale.

The NEP scale consists of 15 items covering a large range of statements on man's role in nature and the role of nature itself.[87] The items were offered as a 5-point strongly agree/disagree scale with a midpoint of "both agree/disagree." Due to the one-dimensionality of the data, we constructed a single additive scale from the 15 NEP items for both groups (see Grendstad, 1999). The minimum score on this scale is 0. It indicates a rejection of the new ecological paradigm. The maximum score is 1. It indicates strong agreement with the new ecological paradigm.

On the NEP scale, the general population has a score of 0.67 and organized environmentalists have a score of 0.77. Unsurprisingly, organized environmentalists hold a more ecological view on nature than what the general population does. Although this difference is significant and in the expected direction, the difference between the two groups cannot be considered overwhelming. Measured by the NEP, the Norwegian environmental movement is quite homogenous.[88] Therefore, the NEP scale distinguishes well between the general population and the organized environmentalists, in the sense that organized environmentalists are more pro-ecological than the general population, but it fails to distinguish between types of environmental organization.

Eckersley (1992) elaborates on the distinctions between anthropocentrism and ecocentrism. In anthropocentrism, man is the aim of history and the end point of evolution, with the right and obligation to manage and control nature's resources. In ecocentrism, all life-forms are regarded as equal, interdependent, and of inviolable intrinsic worth. Consequently, humans represent only one life-form of many. The environmental problems stem from humans overestimating their own importance and neglecting the obligation of empathy and caution toward other species.

---

[87] The 15 items are expected to cluster around 5 subdimensions. (1) Limits to Growth: We are approaching the limit of the number of people that the Earth can support; the Earth has plenty of natural resources if we just learn to develop them; the Earth is like a spaceship with only limited room and resources. (2) Antianthropocentrism: Humans have the right to modify the natural environment to suit their needs; plants and animals have as much right as humans to exist; humans were meant to rule over the rest of nature. (3) Balance of Nature: When humans interfere with nature, it often produces disastrous consequences; the balance of nature is strong enough to cope with the impacts of modern industrial nations; the balance of nature is very delicate and easily upset. (4) Rejection of Exemptionalism: Human ingenuity will ensure that we do not make the Earth unliveable; despite our special abilities, humans are still subject to the laws of nature; humans will eventually learn enough about how nature works to be able to control it. (5) Possibility of an Eco-crisis: Humans are severely abusing the environment; the so-called "ecological crisis" facing humankind has been greatly exaggerated; if things continue on their present course, we will soon experience a major ecological catastrophe.

[88] There are only minor differences between the scores of the different types of organization. Members of new organizations are above average (the means being 0.78), members of old core organizations are at the average (0.77), whereas old noncore members are below average (0.75).

Along the anthropocentrism–ecocentrism spectrum, Eckersley identifies a total of seven subtypes: resource conservation (that nature is valuable primarily when it can be used to satisfy human needs), human welfare ecology (that nature is a necessary material resource for humans and a necessity for health and the social environment), preservationism (that we must protect magnificent landscapes because they satisfy humans' need for enriching experiences in nature), animal liberation (that it is not right to use animals for medical testing even if it might save human lives), autopoietic intrinsic value theory (that all ecological systems, however small and insignificant, have a right to exist), transpersonal ecology (that all human beings must increase their self-awareness so that they can feel at one with all living creatures), and ecofeminism (that women, as opposed to men, have an experiential background that creates greater understanding for the relations in nature).[89]

Although Eckersley's work is often philosophical and theoretical, our approach here is one of plain empirical application. To the best of our knowledge, we are not aware of other such similar empirical applications of her work. Here, Eckersley's perspective is used in three ways: through a anthropocentrism–eco-centrism self-placement scale, through a litmus test on the issues of human population numbers and wilderness preservation through which adherents of anthropocentrism and ecocentrism are separated from one another (Eckersley, 1992, p. 29), and through a set of seven items, each of which relates to the seven subcategories (Grendstad & Wollebæk, 1998).

First, respondents were asked to indicate, on a 7-point self-placement scale, their own position on the centrality of humans in the ecological system: Whether humans are at the center of the ecological system (i.e., anthropocentrism) or merely a small part of the ecological system (i.e., ecocentrism). The scale was transformed and runs between 0 (anthropocentrism) and 1 (ecocentrism).

The result indicates that the general public and the organized environmentalists tend toward ecocentrism, as both means (0.61 and 0.69, respectively) are larger than the midpoint value of 0.50. The response patterns also indicate a bimodal distribution—that the distribution has two humps—for both groups (see Figure 5.4).[90] There are no significant differences between the two samples at the anthropocentric pole. The general population outnumbers the environmentalists at the anthropocentric half of the scale. Toward the ecocentric pole, the general population is surpassed. One-third of organized environmentalists support the extreme position that humans are merely a small part of the ecological system.[91]

---

[89] The three subtypes transpersonal ecology, autopoietic intrinsic value theory, and ecofeminism are sometimes linked to a general concept of ecocentrism. Here, we will treat all seven subtypes on the same analytical level.

[90] There were no significant differences of means between the general population and organized environmentalists ($p < 0.05$).

[91] There were no significant mean differences among the organizational types on this scale.

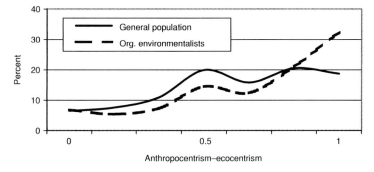

FIGURE 5.4. Anthropocentrism–ecocentrism.
$N_{GP} = 988$, $N_{OE} = 2038$. The curves have been smoothed.

Second, Eckersley (1992, p. 29) offers a litmus test meant to distinguish deci-sively between anthropocentrists and ecocentrists. The test consists of two ques-tions. One question concerns whether to save pristine nature due to the interest of mankind, or regardless of the interest of mankind. The other question concerns how to save the environment: either by limiting population growth, or by decreas-ing, in absolute numbers, the size of the population. The first option for both questions is supposed to identify anthropocentrists. The second option for both questions is supposed to identify ecocentrists. The response pattern allows the respondents to be classified as anthropocentrists, ecocentrists, or as a mix of the two (see Table 5.1).

The analysis showed that the great majority of the general population (77%) and organized environmentalists (89%) chose the ecocentric position of preserv-ing wilderness regardless of mankind. It also showed that a great majority of the general population (89%) and organized environmentalists (81%) chose the anthropocentric position of saving the environment by limiting population growth. The outcome of these choices is that most respondents in the two groups combine an ecocentric and an anthropocentric position.

Concerning the pure categories, the two groups are mirror images of one another. The general population has 1 in 5 pure anthropocentrists as opposed

TABLE 5.1. A classification of anthropocentrists and ecocentrists (in percent).

		General population		Organized environmentalists	
		Population growth			
		A	E	A	E
Wilderness preservation	A	21	3	9	2
	E	68	9	72	17
		101		100	

Note: A: anthropocentrism; E: ecocentrism.
$N_{GP} = 965$, $N_{OEs} = 1982$.

to 1 in 10 for the organized environmentalists. Organized environmentalists have roughly 1 in 5 pure ecocentrists as opposed to almost 1 in 10 for the general population. The main observation is that the litmus test only classifies 30% of the general population and 26% of the organized environmentalists into pure anthropocentrists and ecocentrists. These results suggest that the litmus test does not perform well in separating anthropocentrists and ecocentrists from one another.[92]

The contradictory empirical outcomes of the self-placement and litmus tests might stem from methodological design. The self-placement procedure polarizes two views on a rating scale. It allows respondents to fine-tune their answers along a general and undemanding dimension, along which there are no restrictions to overselling idealism. In contrast, the litmus test forces respondents to make a decision between two pairs of explicit statements. Then it becomes a position too extreme for most people to reduce the population number, but protection of nature that is not directly usable for humans is a much wider appeal.

The frequency of pure ecocentrists and anthropocentrists vary across the four organizational types. New core organizations have the largest number of pure ecocentrists (19%), followed closely by old core (18%) and old noncore (17%), and, finally, new noncore organizations (14%). New noncore organizations have few pure anthropocentrists (5%). Core organizations have twice this number (10 % in both) and there are three times as many in old noncore organizations (15%). Taken together, this indicates that the largest difference among organizational types is as to how many pure anthropocentrists there are in the organizations, not the number of ecocentrists.

Third, seven items, each intended to measure one of Eckersley's seven subtypes, were offered to all of the respondents.[93] A low score indicates disagreement and a high score indicates agreement with the item. The subtypes are ordered according to their face validity relating to our understanding of Eckersley's anthropocentric–ecocentric spectrum (see Figure 5.5).

The relative agreements with four of the concepts—transpersonal ecology, autopoietic intrinsic value theory, preservationism, and human welfare ecology—indicate small differences between the general population and environmental

---

[92] One reason for the lack of performance of the test might be that these two issues are simply too independent of one another: How a respondent answers the first question is not related to how the respondent answers the second question. Another reason is that the litmus test might be inadequate. Wilderness preservation can only become a controversial issue when there is wilderness to preserve (Dryzek, 1993). However, this should make it a controversial question in Norway. Finally, the two questions might be too theoretically advanced, not only for the general population but, as the analysis indicates, even for the rank and file of environmental organizations.

[93] The items were offered as a 5-point strongly agree/disagree scale, with a midpoint of "both agree/disagree" and a "don't know" option (which was set to missing). Seven additional items were offered to the organized environmentalists only, but these items are not included here (see Grendstad & Wollebæk, 1998).

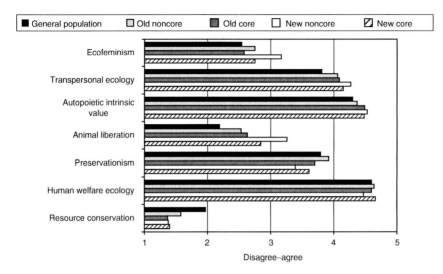

FIGURE 5.5. Subtypes of anthropocentrism and ecocentrism (means). $N_{\mathrm{GP}} = 929$ to 990, $N_{\mathrm{OE}} = 1865$ to 2023.

groups.[94] These response patterns indicate that aspects of anthropocentric and ecocentric concerns generate little if any differences even within our four types. The pattern of greater disagreement found within ecofeminism and animal rights, however, indicate greater contestedness. The contestedness and variation stem in part from the fact that the issues of ecofeminism and animal rights also address directly two of the environmental groups in the new noncore typology, namely NOAH—for animal rights and Women–Environment–Development. However, we also see that the issue of animal liberation sets the general population apart from the environmentalists. In this framework, therefore, animal liberation seems to discriminate the most between the general population and organized environmentalists, with the latter groups being more willing to protect animals even when medical experiments on animals can help to save humans (see Chapter 7). Only on the issue of resource conservation does the general population reach a greater agreement than any of the environmental types. Knowing that Eckersly (1992, p. 35) expected resource conservation to be "the least controversial stream of modern environmentalism", the general disagreement with the issue might stem not only from too strong a phrasing of the item but also from the fact that it is too extreme to claim that nature is valuable primarily when it can be used to satisfy human needs. The view on nature within the antropocentric–ecocentric dimension could also be expressed through two other crucial positions concerning aspects of life: the role of technology and religion in society.

_____

[94] There are significant differences between the general population and organized environmentalists on all items ($p < 0.05$) except on Human Welfare Ecology.

Technology and Religion

Technology is a mixed blessing. On the one hand, it offers solutions to many of the environmental problems that the world is facing, such as reduced pollution and cleaner energy sources through solar energy and fuel cells. On the other hand, many environmental problems are caused by, or intimately linked with, technological developments, such as car, ship, and aircraft transportation. Polluting technologies and heavy industries are often linked to the ideology of economic growth.

Technology is also a question of scale. Large-scale technologies, such as nuclear power plants, seem to be less wanted among environmentalists because such technologies often are linked with big business and industrial complexes. Small-scale technologies, such as personal computers, can be readily applied to improve the running of organizations. The Internet is an inexpensive way to connect to like-minded souls elsewhere. Thus, the environmental movement is trapped in a dilemma in which technological innovation can be regarded both as a driving force of environmental problems and a savior and the solution to the same problems (Coleman, 1994). In order to develop nonpolluting technologies, one has to embrace the innovation that can make this technology possible (Lewis, 1992).

Our survey contains three questions on technology: Whether humans control technology, or vice versa? Whether one understands more or less of natural sciences and technology compared to others? Whether humans will mostly benefit or mostly suffer from technological progress over the next 20 years? Each question was transformed into scales running from 0 to 1. High scores indicate that humans control technology, that humans will mostly benefit from technology, and that one understands more of science and technology than others.

Our analyses show that, on average, organized environmentalists cannot be distinguished from the general population on whether humans control technology or vice versa (see Figure 5.6). We fail to find either a distinct technophobic or technophile position among the vanguards of environmentalism. Perhaps technophobia or technophilia cancel one another out by the item's indistinction of large and small technologies. On the question of one's understanding of natural science and technology, organized environmentalists, when compared to the general population, clearly think that they understand more of such issues than the population at large. However, because we are unable to test such understanding precisely, another way to put it is that the general population has less confidence in such knowledge than organized environmentalists. On the question on technology's future benefit or harm, all groups seem to opt for the Janus-faced solution that it contains both.

The analyses show that there are small differences between the types of environmental organization and between organizational types and the general population.[95] Members of new noncore organizations, however, score lowest on all questions compared to the other three types. They are more pessimistic as to

---

[95] There are significant differences ($p < 0.05$) between the general population and organizational types on all issues but the question on "control."

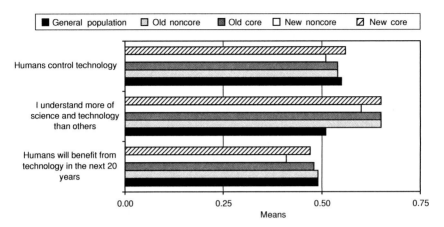

FIGURE 5.6. Technology.
N_{GP} = 978/821/848, N_{OE} = 2028/1836/1844.

humans' control over technology, less confident in own understanding of science and technology, and more cynical about the blessings of technology. Despite the general consensus across the groups on these issues, the results also convey the contestedness of technology. To be precise, the contestedness lies in the social fabric, as we briefly touched on when we discussed grid–group theory earlier, and the way that an issue, event, or contraption, in a sense, is employed by social actors.

This is no less true when we move into another crucial dimension of modern life: religion. In the book of Genesis (1:28), God tells Adam and Eve to "be fruitful, and multiply, and replenish the earth, and subdue it: and have dominion of the fish of the sea, and over the fowl of the air, and over every living thing that moveth upon the earth." Among others, this quote provides the basis for the White (1967) thesis that the Christian dominion doctrine ("the most anthropocentric religion man has seen") fosters negative environmental attitudes and outcomes. Empirical research has offered inconclusive evidence as to the effect of religious attitudes and behaviors on environmental attitudes and behaviors. On balance, however, the White thesis does not receive strong support (Guth, Green, Kellstedt, & Smidt, 1995; Kanagy & Nelsen, 1995; Kanagy & Willits, 1993; Wolkomir, Futreal, Woodrum, & Hoban, 1997). On the other hand one can find that "new religion" or "new-age religion" are linked to environmental concern. New-age religion is less personified and less conventional. Adherents often hold views like reincarnation and the benignness of human nature. Regardless of prefixes, religiosity might indicate a certain awareness or respect for life, making religious people more inclined to become environmentalists.[96] Our interest here

---

[96] In Norway, Botvar (1996) found a positive relationship between all forms of religion and environmentalism. The link, he claimed, is partly found in the degree of frugality advocated by those who claim themselves to be Christians. Frugality, in turn, is a type of behavior that overlaps with conventional measures of environmental behaviors (see, e.g., Hallin, 1995).

is how different measures of religiosity vary across the groups we study. We employ questions on the existence of God, reincarnation, and characteristics of human nature. These questions tap traditional as well as new-age religion.

First, we asked about beliefs in God (see Figure 5.7). Roughly a total of 50% of respondents in all groups expressed that a supreme being exists or that even though in doubt, they believe in God. Members of environmental organizations tended to opt for the existence of a supreme being, whereas the general population more frequently responded that they believe in God. For the general population, roughly 23% said they do not believe in God or are ignorant about God's existence. For both types of core organizations, this percentage is 38%; for new noncore organizations, it is 30%; for old noncore organizations, it is 28%. Hence, the general population is less atheistic and agnostic than organized environmentalists. Members of core organizations are more atheistic and agnostic than members of noncore organizations.

Second, the survey also contained a question on whether man has lived previous lives. This question is a predictor of new-age religion (Donahue, 1993). On this issue, members of old organizations reject reincarnation the most, whereas members of new organizations are more inclined to endorse reincarnation. Members of old core organizations are most skeptical, whereas members of new noncore organizations agree most strongly. However, all organizational types scored below the midpoint of the scale, indicating a general skepticism of reincarnation.[97]

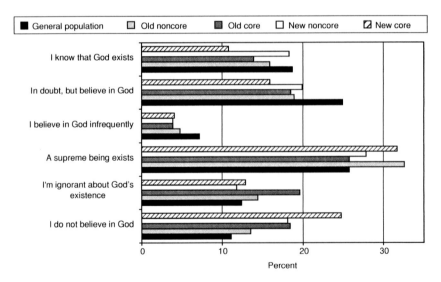

FIGURE 5.7. Religious beliefs.
$N_{GP} = 994$, $N_{OE} = 2038$.

---

[97] This question was posed to organized environmentalists only. On a 0–1 scale, where a high score indicates agreement with reincarnation, the means for the organizational types are as follows: old core, 0.25; old noncore, 0.28; new core, 0.32; new noncore, 0.37.

Third, respondents were asked whether they think humans are basically good or evil. This question might single out adherents of old religion in the Lutheran sense by permitting people to reject the notion of original sin. On a scale that runs from 0 to 1, high scores are associated with humans being basically good. The general response to this question, across all groups, was a bias toward humans being more good than evil. Both the general population mean and the environmentalist mean are 0.69.[98]

In general, organized environmentalists tended to be somewhat less religious than the general population when measured as belief in God. Traditional measures of religion show that the general population believes in a personal God, whereas organized environmentalists more frequently either reject, or are ignorant of, God. Measures related to less traditional religion indicate that organized environmentalists reject new forms of religion such as reincarnation less than the general population.

Conclusion

The environmental concerns and political attitudes of Norwegian organized environmentalists are diverse. Environmentalism can be studied along more than one dimension. However, these dimensions might not always be independent of one another. Environmentalism is not independent of the traditional left–right dimension. Compared to the general population, organized environmentalists have a certain leftist leaning in that they endorse left-wing parties and turn away from parties strongly associated with economic growth. Thus, there is some merit to the "watermelon hypothesis": Once the green jacket is removed, environmentalists are all red on the inside. Compared with the general population, organized environmentalists hold more ecological beliefs, are more postmaterialist, and are egalitarian.

However, when we compare organized environmentalists and the general population, we are struck by the absence of large and fundamental differences. Although organized environmentalists sometimes differ in a radical and ecological direction, they fail to be distinguished from the general population in any profound way. This is not what one should expect if environmental organizations were part of an autonomous subgroup of society that organizes itself in opposition to the state, which is a view that we find in the Dryzek ideal of an autonomous civil society (see Chapter 2). On questions such as postmaterialist values, the supremacy of humans or nature, views on technology, or adherence to religion there seem to be no fundamental differences between the general population and organized environmentalists.

So far we have failed to detect strong and significant differences among the various types of environmental organization. However, is this really the case?

---

[98] For adherents within each organizational type, the means run from 0.68 to 0.71. No significant differences are found among any group. Originally, the scale's range consisted of seven points.

Because there is a change in the way new organizations operate and there is a difference in what issues environmental organizations pursue (see Chapter 3), we end this chapter with an analysis specifically aimed at finding differences among the four organizational types. For this purpose, we cast the variables from the previous analyses in this chapter into a discriminant analysis. The object is to identify which, if any, of the attitudinal variables, in isolation or in combination, are capable of discriminating the most among the four types of environmental organization.[99] After an initial and explorative analysis using all relevant variables, we recast the four variables capable of discriminating most among the four groups into a final analysis to facilitate interpretation of differences among organizational types. These variables are preservationism, the New Ecological Paradigm Scale, individualism, and the belief that people have lived previous lives (reincarnation) (see Table 5.2).

The first discriminant function picks up Eckersley's preservationism (the protection of magnificent landscapes to satisfy humans' need for enriching experiences in nature) and, inversely, Dunlap's new ecological paradigm scale

TABLE 5.2. Political and environmental beliefs. Discriminant analysis.

	Function 1	Function 2	Function 3
Preservationism	0.88	0.03	−0.06
New ecological paradigm	−0.49	0.52	−0.06
Individualism	0.32	0.78	0.10
People lived previous lives	−0.09	0.12	0.98
Eigenvalue	0.04	0.01	0.01
Centroids			
New core	−0.04	0.20	0.04
Old core	0.04	−0.06	−0.13
New noncore	−0.20	−0.01	0.15
Old noncore	0.28	−0.19	−0.03

Note: $N_{OEs} = 1314$.
Functions contain rotated loadings (correlations). Differences between groups are significant at $p < 0.05$.

---

[99] Discriminant analysis (DA) uses a set of metric variables to *maximize differences* between groups (here, types of environmental organization). Variables on a nonmetric scale are excluded (e.g., beliefs in God). DA weights the variables onto one or more discriminant functions (i.e., composite indexes) so that each function maximizes differences *between* groups. Any other weighting of the variables will yield smaller differences among the organizational types. The value to be maximized is the "eigenvalue" (i.e., the ratio of between-group variance to within-group variance). "Group centroid" is each type's mean score on the discriminant functions. A DA is also evaluated on (1) the interpretability of the discriminant functions (through the correlation coefficients when they exceed absolute values of 0.30) and (2) relative positions of group centroids. Because proximity between discriminating variables and the groups to be profiled can bias the outcome of the analysis (see Klecka, 1980; Nunnally & Bernstein, 1994), two variables were removed from the analysis (i.e., "animal liberation" and NOAH—for animal rights, and "ecofeminism" and Women–Environment–Development).

(a rejection of human exemptionalism and a belief that any progress must pay careful attention to the limitations found in nature itself and the ways in which humans interact with nature). This function polarizes new and old noncore organizations. New noncore organizations are identified with the New Ecological Paradigm Scale (and antipreservationism), whereas old noncore organizations are identified with preservationism (and the Dominant Social Paradigm). Core organizations of both types fail to be distinguished along this cleavage. What we see here is a kind of symmetry between members' attitudes and the goals of the organizations (see Strømsnes, 2001, and Chapter 3 of this volume).

The second discriminant function combines the New Ecological Paradigm Scale and individualism's equal opportunity. We coin this brand of environmentalism "ecological individualism." It polarizes old noncore and new core organizations. Ecological individualism is a term that fits well for members of new core organizations. Again, we see a symmetry between members' attitudes and organizational goals. Members of new core organizations are action-oriented and emphasize that the battle of environmentalism can only be won through a knowledgeable elite. It makes sense that this view is contrasted to old noncore organizations whose traditional emphasis on recreation and resource extraction have been on what nature can do for you.

The third discriminant function picks up new-age beliefs: that people have lived previous lives. The function polarizes new noncore from old core organizations. This indicates that members of new noncore organizations are more receptive to alternative ways of thinking relating to spirituality, awareness, and holism. It is probably part of the new modern streams of ideas that are broader than being environmental only, but in which environmental concern is an important part.

In our analyses, we have presumed that being positioned at the core is a goal for an environmental organization. Therefore, it is interesting to note that the function with the most discriminating power between organizational types (i.e., preservationism versus ecological beliefs) does not affect what we have defined as core types of environmental organization. On the most important function, the members of core organizations are "center-oriented" in the sense that they fail to hold extreme views. Here, the bottom line is that "coreness" is associated with "center-oriented" or moderate environmentalism. The analyses have also shown that the most radical environmental views are found in new noncore organizations. In old noncore organizations, environmentalists are not at all radical. The center category is held by core organizations that we had hypothesized as holding moderate views on environmentalism. *Being at the core of the Norwegian environmental movement implies simply a center-oriented and rather moderate form of environmentalism.* These results, we believe, support both the state-friendly society anomaly and the local community anomaly. Both anomalies direct the environmental movement in a pragmatic direction. These results are in line with the Dryzek study's (2003) overall understanding of the ideological position of Norwegian environmentalists being within the mainstream of the national political culture. However, we disagree with Dryzek on the causes of this mainstream position.

Chapter 6
Political and Environmental Behavior

Introduction

Environmental beliefs and attitudes can make a difference. They can lead people to comprehend to a greater degree the nuts and bolts of the relationship between society and nature. Environmental beliefs and attitudes also hold out the promise of being a behavioral trigger. So when these beliefs and attitudes are translated into behavior, environmentalism becomes real in its consequences. However, researchers have also shown that for some people, environmental behavior might be unrelated to environmental beliefs and attitudes. For example, the acts of recycling household waste might simply be acts of frugality stemming from a general idea of resource conservation or even from childhood socialization during more difficult times (Hallin, 1995). Also, being out in nature might take on another meaning, knowing that this is simply something that farmers and foresters do.

When environmentalism becomes politicized, environmental behavior might also be seen as variations of political behavior. We approach the analysis of environmental behaviors from a political science approach. First, we look at conventional and unconventional political activity. Second, we analyze environmental behavior more explicitly. Although we expect organized environmentalists to outperform the general public regarding environmental behavior, we are also interested in the analyses of whether organized environmentalists are more or less active outside of the organization to which they belong (see also Chapter 8). Third, anthropologists have taught us about purity and danger. We pick up on these concepts and apply them to the environmentalists' behavior. Here, we study the degree to which organized environmentalists hold different conceptions of purity and danger compared to the general population. Finally, we turn to the tenuous relationship between attitude and behavior.

Political Behavior

Although the distinction might have become somewhat blurred over the years, we distinguish between "conventional" and "unconventional" political behavior. Conventional political behavior is strongly linked to areas like party membership, voting, and running for office. Unconventional political behavior, on the other hand, is linked more to different ways of fielding protest like boycotts and civil disobedience (see Dalton, 1988; Rootes, 2003; Raaum, 1999; Strømsnes, 2003). Our survey informs us that 55% of the general population is fairly or very interested in politics. The figure for organized environmentalists is 74%. Even if the figure for environmentalists is significantly higher compared with the population at large, one should perhaps have expected a greater difference between the two groups in such a clearly political field.

Membership in environmental organizations is not only a political act in itself; it is more of a political act than being a member in most other types of voluntary organization because participation in such voluntary organizations, at least in accordance with theory, is more outside of mainstream politics, is more explicit on ideology, and often entails a greater commitment to societal change. It has been shown in contexts other than the one we analyze here that membership in voluntary organizations augment societal commitment, such as various forms of political activity (Wollebæk, 2000). However, most voluntary organizations are evidently less political than those within the environmental field that we are addressing here. In the following analyses, we assume that the difference in environmentalists' political behavior are of the same proportions when stacked against the general population. We also assume that differences in activities between the general population and the organized environmentalists will increase when unconventional activities are observed.

The results of analyzing conventional political behavior show that, with the exception of voting, organized environmentalists participate more within conventional political behavior than do the general population (see Table 6.1). Approximately 90% of all respondents answered that they usually voted in local and national elections.[100] In local elections, less than half of the general population has made changes to the ballots, whereas more than half of all environmentalists have made such changes. Among the general population, one in four has attended a local council and committee meeting, whereas 15% have tried to influence politicians directly. For organized environmentalists, both of these numbers are close to 40%. Fourteen percent of the general population reported that they are members of a political party, whereas only 11% have held some type of local political office.

Among organized environmentalists, 20% have joined a party and approximately 15% have held a political office (see Chapter 5). This indicates that

---

[100] We assume these figures to be slightly inflated because respondents tend to overreport behavior that is commendable and considered a citizen's basic duty.

TABLE 6.1. Conventional political behavior; percent "yes".

	GP	ONC	OC	NNC	NC
Do you usually vote in local government or national parliamentary elections?[a]	92	96	90	88	95
In local elections you have the option of changing party or group lists. You may either cross out names or add new ones. Have you ever changed the party ticket in this way?[b]	44	53	47	51	54
Local government council meetings and some local committee meetings are open to the public. Have you ever attended such meetings?[b]	26	36	34	40	34
Have you ever tried to influence political decisions through direct contact with politicians?[b]	15	39	36	37	37
Are you a member of a political party?	14	24	19	20	16
Have you ever, or are you currently serving on the local council executive, the regional parliament or any other local or regional government body?[b]	11	17	15	18	12

[a] $p > 0.05$ between the general population and organized environmentalists.
[b] $p > 0.05$ between types of organization.
$N_{GP} = 996–999$, $N_{OE} = 2031–2044$.
Note: GP = general population; ONC = old noncore; OC = old core; NNC = new noncore; NC = new core.

environmentalists have not abandoned political parties in favor of other types of voluntary organization, but that environmentalists more than the general population are incorporated in conventional politics. This result is not surprising in a state-friendly society. Except for voting and membership in political parties, differences among organizational types cannot be found. Further analysis shows that the six measures of conventional political activity that we have used in Table 6.1 reflect a single underlying dimension of political behavior for both the general population and organized environmentalists. This shows that conventional political behavior is unidimensional and additive and that there is no pattern of participation among environmentalists different from that of the general population.[101]

For unconventional political behavior, we find it fruitful to analyze whether or not organized environmentalists act in the capacity of being members of their organization. The six behaviors, within a time frame of 3 years, are as follows:

---

[101] Principal component analysis, varimax rotation. A weak second factor consisted of the variable "voting in elections" and was joined by the variable "changing party lists" on the rotated second factor. The same results as those obtained here were also found by Wollebæk and colleagues (2000) for the voluntary sector in general.

signed a petition; taken part in a protest march or a local protest action; boycotted goods; contributed financially to a protest; written articles for a newspaper or periodical on current issues; taken part in illegal protests (civil disobedience) in order to prevent the carrying out of governmental plans. If the organization in which an environmentalist is a member organizes the activity, then the environmentalist acts as a member. If not, the environmentalist participates outside of, or independently of, the organization. This distinction indicates the degree to which an organization consumes the political life of its members. Relatedly, the distinction also reflects the extent to which members are thoroughly socialized by the organization.

First, we analyze the structure of environmentalists' unconventional political behavior as members of their organization. An explorative principal component analysis of all six measures yields one robust dimension across all four organizational types.[102] We therefore conclude that unconventional political behavior consists of one main dimension only. Because there are no independent domains or cross-cutting dimensions within unconventional behavior, we can measure unconventional political behavior in terms of more or less behavior along this dimension.

The results show that organized environmentalists outperform the general population on all types of unconventional political behavior (see Table 6.2). The results also show that environmentalists are more active *outside* of their own organization than within their organization. The largest differences among organizational types are found when one participates as a member of an environmental organization. The most passive members are found in old noncore organizations. Members of new noncore organizations are particularly active in signing petitions, protest marching, boycotting goods, and writing articles. Old core members also sign petitions and participate in protest marches. Along with new core members, they also contribute monetarily most frequently to a protest campaign.

Again, we use discriminant analysis to find out which type or types of unconventional behavior within one's organization distinguish the most among the organizational types. The analyses reveal that boycotting of goods is the main difference among the organizational types. It separates members of new noncore organizations (often boycott) from members of both types of old organization (seldom boycott). The secondary difference is funding of political protests. This type of behavior separates members of core organizations (often contributes) from members of noncore organizations (seldom contributes). The marginal difference is found in signing a petition. This type of behavior separates members of old noncore organizations (infrequent) from members of new noncore organizations (frequent).

With the exception of voting, organized environmentalists outperform the general population in all forms of conventional and unconventional political behavior. We also observed that the members' political activity is not only limited to the organization to which they are members. Members are sometimes more active

---

[102] A second component was discerned when requested, but it did not add significantly to the first unrotated component (dimension) to which all variables contributed.

TABLE 6.2. Unconventional political behavior; percent "yes".

			Participation independent of organization				Participation as member of organization			
	Freq.	GP	ONC	OC	NNC	NC	ONC	OC	NNC	NC
Signed a petition[a]	1–2	31	40	40	40	44	20	35	31	30
	3+	8	21	26	29	25	5	13	20	11
Protest march	1–2	13	22	22	27	19	4	14	13	10
	3+	2	10	14	11	17	1	5	5	2
Boycotted goods	1–2	21	23	30	26	27	18	26	20	24
	3+	12	32	29	42	38	18	24	45	31
Financial contribution[a]	1–2	32	28	34	31	33	30	40	29	40
	3+	21	32	37	34	34	11	24	22	23
Written articles	1–2	6	13	13	19	17	8	8	13	7
	3+	2	8	7	13	12	3	4	7	6
Civil disobedience[a]	1–2	2	3	4	5	5	1	3	3	1
	3+	0	0	2	2	2	0	1	2	2

[a] $p > 0.05$ between types of organization (i.e., participation independent of the organization).
$N_{GP} = 975–989$, $N_{OEs, independent} = 1809–1851$, $N_{OEs\ as\ members} = 1943–1964$.
Note: GP = general population; ONC = old noncore; OC = old core; NNC = new noncore; NC = new core.
Freq. refers to the number of times the behavior in question has occurred.

outside, than inside, the organization to which they belong. This suggests that membership is but one of several ways in which organized environmentalists participate in politics. Being a member of an environmental organization augments unconventional political activism.

Environmental Behavior

In the survey, we offered the respondents a list of 16 items of private and individual environmental behaviors. For each behavior, we asked whether it was performed always, sometimes, or never.[103] After an exploration of the relationship

---

[103] There are two considerations linked to environmental behavior—or political behavior for that matter—that we will not address here. First and as we already touched upon in the introduction to this chapter, a type of behavior cannot be environmental in its own right. Cutting back car use might be justified with reference to personal health if there is a need for more physical exercise or it can be justified with reference to one's personal economy if gasoline costs an arm and a leg. However, it might also be justified environmentally by reducing, however minuscule, the amount of carbon dioxide (CO_2) released into the atmosphere, thereby lessening the greenhouse effect. Second, environmental behaviors are, as economists like to tell us, linked to costs. Costs can be measured in the amount of resources that are needed to carry out the behavior. Being out in nature is fairly easy if one lives well outside urban areas, whereas avoid using a car might be difficult if one needs to travel longer distances where public transportation or other means are unavailable (see, e.g., Olli, Grendstad, & Wollebæk, 2001).

among these items, we were able to order the 16 behavioral items by 5 general categories: *responsible consumerism*: to choose environmental seal of approval, to use unbleached paper, to choose recycled products, to avoid unnecessary packaging, and to avoid chlorine-based products; *resource conservation*: to repair things, to avoid disposable products, to give away used clothing, and to avoid using a car; *use of nature*: to be out in nature and to harvest fruits of nature; *antitoxic*: to avoid toxic products in the yard, and to buy food that is produced without the use of herbicides, pesticides, or chemicals; *waste handling*: to make compost, to gather problem waste, and to sort household waste for recycling (see Olli et al., 2001). For each of the 16 items, the respondents received a score of 1 if the act is "always" performed, a score of 0.5 if the act is performed "sometimes," and a score of 0 when the act is never performed. Therefore, one unit corresponds to one environmental act. Again, we assume that organized environmentalists have the home-field advantage and will, therefore, outperform the general population.

Organized environmentalists report higher levels of environmental behaviors than the general population across all categories (see Table 6.3). On the other hand, the general population is not doing so bad either. It has scores above the middle value of the scale on all constructs except waste handling, which is known to be a type of behavior that entails both motivation and costs. The high scores of the general population across the types of behavior can be linked to the local community perspective in which there is a strong tradition of not inflicting harm on nature and for acting in a sustainable manner (see also Chapter 7). The results also show that members of new noncore organizations score highest on all constructs except "use of nature," where members of old noncore organizations report higher activity. It is also clear that members of new organizations behave more environmentally friendly than do members of old organizations. All in all, we see a symmetry between the behavior of members and the features of the organizations of which they are members (Strømsnes, 2001).

A discriminant analysis reveals that responsible consumerism is the behavioral category that most effectively discriminates among types of environmental organization. Responsible consumerism separates members of new noncore organizations (high score) from members of old noncore organizations (low score).

TABLE 6.3. Environmental behavior (means).

	GP	ONC	OC	NNC	NC
Use of nature (0–2)	1.5	1.8	1.6	1.6	1.6
Responsible consumerism (0–5)	3.4	3.9	4.0	4.3	4.1
Antitoxic (0–2)	1.2	1.4	1.4	1.5	1.4
Resource conservation (0–4)	2.2	2.7	2.8	2.9	2.8
Waste handling (0–3)	1.4	2.2	2.0	2.2	2.1

Note: The 10 analyses of variance are $p < 0.05$.
$N_{GP} = 950–774$, $N_{OE} = 1618–1988$.
The parentheses show the range of the general behaviors. The behaviors are sorted according to maximum relative score for the general population.

The second best discriminator is use of nature, which separates old noncore members (high score) from members of new core organizations (low score). Again, these findings point in the direction of symmetry between organizational features and member activity. It seems that the organizations' emphasis on either conservation or pollution, to some extent, reflects on the behavior of their members (see Chapter 3).

In addition to their personal environmental behaviors, organized environmentalists can also make temporal and monetary contributions. In the survey, the organized environmentalists were asked to report how much time (hours per month) and how much money (per year) they spend on environmental work.[104] The data show that members of new noncore organizations spend on average 5.6 hours a month, whereas new core environmentalists spend 4.7 hours a month on environmental work. For members in old core and old noncore organizations, the averages are 3.3 and 3.6 hours, respectively. Thus members of new organizations contribute the most time to environmental work. As to annual monetary contributions, the score for new core members is NOK 956, and NOK 903 for old noncore members. For new noncore members, the number is NOK 767, and NOK 736 for old core members. Thus, the most money contributed to environmental work is found in new core organizations and in old noncore organizations. Members of old core organizations make the least temporal and monetary contributions. When we compare this with the voluntary sector at large, these figures are not specifically high. Although the monetary contributions are about the same amount as found in many other types of organization, the average member of any voluntary organization give more than 10 hours a month to voluntary work. Still, these figures are dwarfed by what can be observed in religious organizations, as well as in social and welfare organizations (see Wollebæk et al., 2000, especially Chapter 2).

We also tested whether there exists a trade-off between monetary and temporal contributions. Limitations on one's personal time might lead to greater monetary contributions, or vice versa. Analysis of data shows that this is not the case. The Pearson's correlation coefficients, which measure degree of linear association between variables, range from 0.30 to 0.50 for the different organizational types ($p < 0.01$). That the correlation coefficients between monetary and temporal contributions are strong and positive for all organizational types indicate that temporal and monetary contributions are additive. Those who participate little, also pay little. Those who participate the most, also pay the most. This positive association between temporal and monetary contributions is also found in the voluntary sector in general (Wollebæk et al., 2000).

---

[104] The variables of temporal and monetary contribution values were somewhat skewed. In order not to bias the analyses, we excluded all values one standard deviation above the means (i.e., scores exceeding 34 hours of environmental work per month and monetary contributions exceeding NOK 11,000 per year (NOK 700 approximates $100). This exclusion amounted to 4.2% and 0,8% of the cases, respectively.

We also included in our survey a question on how active the environmentalists consider themselves to be in the organization. Here, we consider the respondents active if they, as members of the organization, participate at least a couple of times each year. Less participation leads to them being categorized as passive. Although this criterion might be considered modest, only one in five of all organized environmentalists are considered active by this standard. When our distinction of active/passive is applied within the organizational types, we find a larger share of active members in noncore organizations (29% and 31%) than in core organizations (10% and 14%; see row A in Table 6.4).

The active–passive distinction can also be used to assess temporal contributions (see row B in Table 6.4). Active members of new core and new noncore organizations are, on average, involved in environmental work 10 hours a month (i.e., approximately the same amount as the "mean" member of the voluntary sector at large). In these organizations, passive members work roughly 4 hours a month. These numbers are lower for members of old organizations.[105] Table 6.4 also shows few members of old core organizations being active and that their participation is low. In terms of financial contributions (see row C), there is a consistent pattern that active members donate more money than passive members. Again, we see the clear correlation between temporal and monetary contributions that we observed earlier. Active members both work and pay more than passive members across all types of organization.

Although new organizations often can be considered nondemocratic (by classical standards of participation and involvement) and hence might suppress the importance of member commitment, these figures give us no reason to conclude that members of new organizations are less active than members of older and

TABLE 6.4. Level of activity and temporal/monetary contributions.

	New core		Old core		New noncore		Old noncore	
	Active	Passive	Active	Passive	Active	Passive	Active	Passive
A. Distribution between active and passive (%)	10	90	14	86	29	71	31	69
B. Environmental work per month (average hours)	10	4	7	3	10	4	5	3
C. Monetary contribution per year (average NOK)	1801	853	1162	660	1060	646	1136	777

Note: In rows B and C, questions on time and money do not control for organizational membership (i.e., inside or outside the organization).
Significant differences between the eight groups $p < 0.01$.
$N_{OE} = 1335$–2029.

---

[105] Here, we are unable to determine whether active environmentalists sign up in new organizations, or new organizations produce active environmentalists. New nondemocratic organizations without ordinary members need a higher degree of participation from a small group of activists to run the organization. Without significant contributions from the chosen few, organizational survival could be at risk.

more democratic organizations. Especially, new noncore organizations have many active and committed members. Active members of new core organizations contribute by far the largest annual sum of money.[106] All in all, there seems to be more activity in the new type of organization and less activity in the old type of organization than should be expected from the organizational model we discussed in Chapter 3.

Purity and Danger

In order to distinguish between individuals and between organizations, we can also draw on another perspective that emphasizes that all organizations need some sort of signs, images, beliefs, or behaviors to demarcate them from other organizations. An organization is considered to be more distinct the stronger and less transient the borders are.[107] The members of these organizations might develop symbols of heroes and enemies and, in turn, use these symbols to increase coherence inside the organization and to demarcate themselves from other groups. These symbols and behaviors translate into notions of the purity found inside the organizations in contradistinction to the dangers found in the outside world. An organization's notions of purity and danger might seem irrational to observers outside of the organization, but they might be perfectly functional to those on the inside. Disputes on purity and danger originate in situations when the members of the organization are unable to agree on where the organization and membership end and where the outside world and non-membership begin.

In her elaboration of these phenomena, the British anthropologist Mary Douglas (1966) set out to classify collective notions of purity (Fardon, 1999). For instance, collective rules on what to eat or not, in which Douglas referred her readers to Leviticus (especially Chapters 11–16), relate primarily to social classification systems. Anomalies, according to these classifications, are matter out of place and if not uprooted will undermine the organizations purity and worldview. By way of extension, members of organizations might also reject ideas, beliefs, behaviors, or commodities because they represent or symbolize institutions whose goals, policies, or histories should not be supported by or associated with the value system of the group.

We apply Douglas' ideas of purity and danger to organized environmentalists in an attempt to explore ways in which they differ from the general population on selected issues of consumption and lifestyles (see Table 6.5). Ready-made

---

[106] Given the implausible claim that members of new nondemocratic organizations are close to being irrelevant in a social capital perspective (Putnam, 2000), the results obtained here will be important when we discuss the role of members in new organizations in Chapters 7 and 8.

[107] For a comprehensive discussion on organizational boundaries and organizational autonomy in the voluntary sector, see Tranvik and Selle (2003).

food often comes with a host of additives and preservatives, whereas coffee, alcohol, and tobacco contain toxics.[108] These substances might, for instance, violate the notion of "the body as a temple" or being associated with industries or country of origin that can be denounced by environmentalists. Physical exercise and yoga relate to body and soul, respectively, and can be seen as ways to improve fitness and consciousness. The choice of a vegetarian lifestyle might stem from a rejection of the policy of industrial meat industry (including genetically modified food and/or inhumane treatment or novel feeding of animals) or a moral rejection of eating anything that has a face (because animals, too, are sentient beings and therefore, more generally, should be considered on a par with humans).[109] Being a pacifist might signify an unwillingness to carry an instrument of death or it might signify a rejection of the code of military hierarchies. Feminism might be a commitment to develop communities outside of men's domain or a rejection of masculine values in all facets. Pacifism and feminism are closely linked to new politics and alternative social movements.[110] In conclusion, we expect environmentalists to adopt a "pure" alternative more often than the general population.

Except for consumption of ready-made food, for which no difference between organized environmentalists and the general population is found, the analyses show that there are significant differences between these two major categories on all items of purity and danger that were asked to both groups (see Table 6.5).

TABLE 6.5. Purity and danger; percent "yes".

	GP	ONC	OC	NNC	NC
Does not buy a lot of ready-made food[a]	72	75	68	74	69
Does not smoke	70	88	85	79	75
Physical exercise	49	64	62	57	53
Does not drink coffee[b]	31	34	39	40	37
Teetotalist[b]	21	16	14	18	13
Pacifist	12	30	31	44	37
Vegetarian	2	5	4	19	8
Does not drink Coca Cola[b]	na	47	46	50	44
Does not eat whale meat	na	28	39	52	47
Feminist	na	20	29	39	24
Yoga/meditation[b]	na	10	8	10	10

[a] $p > 0.05$ between the general population and organized environmentalists.
[b] $p > 0.05$ between types of organization.
Note: na = not asked.
$N_{GP} = 948$–1018, $N_{OE} = 1951$–2057.

---

[108] Although we understand that poison is found in the dose, purity prefers absence over gradualism.
[109] Vegetarianism on the basis of food intolerance or vegetables being less expensive than meat are justifications too. However, the latter is not very likely in Norway because Artic agriculture and protection from the world market usually boost the prices of vegetables.
[110] For a study on the link between vegetarianism and feminism, see Adams (1990).

With the exception of teetotalism, organized environmentalists have higher scores on all items, as expected. Most notably, environmentalists, compared to the general population, smoke less, exercise more, drink less coffee, and, to a greater extent, consider themselves pacifists. These figures indicate that organized environmentalists report stronger convictions of purity and danger than does the general population. When comparing the four organizational types, members of new noncore organizations can be considered purer than members of the other types. Again, this observation indicates that we are dealing with an ideological or cognitive reorientation that is broader than environmentalism itself. Less evident, members of noncore organizations report purer lifestyles than do members of core organizations. We find this observation rather interesting because it suggests that environmentalism, in the sense of advancing pure lifestyles, is less related to the environmental coreness. Again, *coreness in the case of unique environmentalism actually signifies being mainstream, as our two anomalies of local community and state-friendliness would lead us to expect.*

The Attitude–Behavior Correspondence

Several studies on attitude–behavior correspondence conclude that there exists a tenuous relationship between the two (e.g., Ajzen & Fishbein, 1977; Holland, Verplanken, & Van Knippenberg, 2002; Kraus, 1995; Pooley & O'Connor, 2000; Stern, 2000). One approach to this problem is by way of the asymmetry between attitudes and behaviors:

> Although for every act there is an implicit or explicit belief to justify the act, the reverse is not true; not every thought, fantasy, image, or argument is reflected in behavior, especially since thoughts often rehearse alternative lines of behavior. The world of behavior, therefore, is smaller than the world of thought; the two worlds are not isomorphic (Lane, in Wildavsky, 1987, p. 18).

This claim should not prevent us from observing attitude–behavior correspondences. However, one should not expect these correspondences to be strong. We end this chapter by linking political and environmental behaviors to corresponding beliefs and attitudes to analyze the extent to which this correspondence can be observed for organized environmentalists and the general population.[111]

We start with an analysis of the relationship between political interest and political behavior. Those with high interest in politics (respondents reporting "very" or "quite interested") are expected to be more politically active than those with low interest in politics (respondents reporting "slightly" or "completely uninterested"). The analyses confirm this expectation (see Table 6.6). Those with high political interest are more politically active than those with low political

---

[111] In this section, we do not break the analyses down by the organizational typology due to compounding intricacy and small N.

TABLE 6.6. Political interest and conventional political activity; percent "yes".

	GP Political interest		OE Political interest	
	Low	High	Low	High
Voting in elections	87	96	83	95
Changed the ballot	33	52	31	58
Meeting attendance	15	35	18	42
Influence political decisions	7	21	16	45
Member of a political party	8	19	7	24
Political offices	4	17	3	20

Note: N_{GP} = 984–986, N_{OE} = 2026–2039

interest. With the exception of voting, organized environmentalists with high interest in politics are more politically active than those in the general population with high political interest. However, it is also the case that those in the general population with high political interest are more politically active than organized environmentalists with low political interest. This is a very interesting finding, but we are unable to decide to what extent this difference is a self-selection effect of being environmentally concerned or whether it is an effect of organizational socialization. The comparison between the general population and organized environmentalists with low political interest shows that differences are nonexistent or surprisingly small. The general population with low political interest vote more frequently than do the organized environmentalists who are politically disinterested. Attempts to influence political decisions are the type of political behavior that, by far, set politically interested environmentalists apart from the politically interested general population.

The pattern of correspondence between political interest and unconventional political activity (see Table 6.7) is similar to that between interest and conventional political activity. However, here it should be noted that politically interested organized environmentalists are more politically active outside than inside the organization of which they are members. The single type of behavior that distinguishes organized environmentalists from the general population, regardless of degree of political interest and membership activation, is the frequency of boycotting goods. The frequency of not buying a specific product is a significant attribute of organized environmentalists.

We also expect to observe a correspondence between environmental beliefs and environmental behavior. For this purpose, we link the New Ecological Paradigm Scale (NEP) to the five types of environmental behavior we identified earlier: responsible consumerism, resource conservation, use of nature, antitoxic, and waste handling. To simplify the analysis, we dichotomize the NEP scale using the mean of the general population as the cutoff point for both groups. A score equal to or less than 0.67 is coded "low," indicating low ecological beliefs, whereas a score exceeding 0.67 is coded "high," indicating an endorsement of ecological beliefs. The analysis shows a general positive correspondence between ecological beliefs and environmentally friendly behavior (see Table 6.8).

TABLE 6.7. Political interest and unconventional political activity; percent "yes".

	Political interest	GP		Participation independent of organization		Participation as member of organization	
		Low	High	Low	High	Low	High
	Freq.						
Signed a petition	1–2	27	34	35	43	24	33
	3+	3	12	15	29	9	14
Protest march	1–2	9	16	18	25	7	13
	3+	0	4	4	16	3	4
Boycotted goods	1–2	19	24	26	28	20	24
	3+	7	17	23	39	25	32
Financial contribution	1–2	29	35	28	33	31	37
	3+	19	23	26	38	18	22
Written articles	1–2	1	10	9	18	6	10
	3+	1	3	3	12	2	6
Civil disobedience[a, b]	1–2	1	2	3	5	2	2
	3+	0	0	1	2	1	1

[a] $p > 0.05$ between the two groups in the general population.
[b] $p > 0.05$ between the two groups in the environmental movement as members.
$N_{GP} = 963-977$, $N_{OE, independent} = 1781-1824$, $N_{OE \text{ as members}} = 1910-1933$.

Unsurprisingly, high scores on the NEP are associated with a higher number of environmental behaviors for both the general population and organized environmentalists. Environmentalists, regardless of their values on the ecological concern, perform environmental behaviors more frequently than do the ecologically concerned or unconcerned general population. The differences between the general population and environmentalists are smallest for the less demanding behavioral type of using nature. The differences between the general population and environmentalists are largest for the more demanding task of waste handling. Being organized boosts your ability or is conducive to act in an environmentally friendly way.

A stronger test of the attitude–behavior correspondence can be obtained by using correlation analyses. In this test, we retain the five types of environmental

TABLE 6.8. Ecological concern and environmental behavior (means).

Ecological concern (NEP)	GP		OEs	
	Low	High	Low	High
Responsible consumerism (0–5)	3.2	3.5	3.8	4.1
Resource conservation (0–4)	2.1	2.3	2.6	2.9
Use of nature (0–2)[a]	1.4	1.5	1.6	1.6
Antitoxic (0–2)	1.1	1.2	1.3	1.5
Waste handling (0–3)[b]	1.4	1.4	2.0	2.1

[a] $p > 0.05$ between the two groups in the environmental movement.
[b] $p > 0.05$ between the two groups in the general population.
Note: $N_{GP} = 711-873$, $N_{OE} = 1509-1848$.

behavior and correlate them with the materialist–postmaterialist index, the anthropocentrism–ecocentrism self-placement scale, and the four cultural biases of grid–group theory: individualism, fatalism, egalitarianism, and hierarchy (see Table 6.9).[112]

The analysis shows that for both the general population and the organized environmentalists, a high degree of individualism is associated with a reduced number of environmental behaviors. On the other hand, a high degree of egalitarian biases is associated with a higher number of environmental behaviors. Where significant, considering oneself an ecocentrist corresponds with environmental behaviors. Holding fatalistic biases is unrelated to environmental behaviors for the general population, but it is negatively associated with environmental behaviors among organized environmentalists. With the exception of "use of nature," being a postmaterialist is surprisingly associated with environmental behaviors for organized environmentalists only. In sum, we see that egalitarianism, ecocentrism, and postmaterialism are mostly positively associated with environmental behavior, whereas individualism and fatalism are negatively associated. More generally, in these analyses of the relationships among environmental concerns, political orientations, and environmental behavior, we observe the expected, albeit tenuous, relationship between attitudes and behavior.

Conclusion

Almost consistently, organized environmentalists are more political active than the general population across all types of conventional and unconventional political behavior. Unsurprisingly, organized environmentalists behave consistently more environmentally friendly than do the general population. Organized environmentalists distance themselves from the general population on selected items

TABLE 6.9. Beliefs, attitudes, and environmental behaviors (Pearson's R).

	GP						OEs					
	PM	A–E	INDI	FATA	EGAL	HIER	PM	A–E	INDI	FATA	EGAL	HIER
Responsible consumerism	na	—	−0.14	—	0.24	—	0.12	0.11	−0.11	−0.06	0.15	−0.10
Resource conservation	na	0.07	−0.16	—	0.18	0.08	0.12	0.08	−0.17	−0.11	0.14	−0.08
Use of nature	na	0.10	−0.09	—	0.13	—	—	—	−0.06	−0.06	—	0.10
Antitoxic	na	—	−0.10	—	0.14	—	0.11	0.10	−0.08	—	0.18	−0.09
Waste handling	na	—	—	—	0.08	—	0.05	—	−0.10	−0.05	0.11	—

Note: na: not asked; —: not statistically significant at 0.05; PM: postmaterialism index; A–E: anthropocentrism–ecocentrism self-placement scale; INDI: individualism index; FATA: fatalism index; EGAL: egalitarian index; HIER: hierarchy index. Pairwise deletion of missing cases, N_{GP} = 682–919, N_{OEs} = 1304–1956.

---

[112] See Chapter 5 for a discussion of these items.

of purity (e.g., more vegetarians) and danger (e.g., fewer smokers). Because these types of behavior also consistently relate to political interest and environmental beliefs, we are confident that our observations are valid and that significant differences between the general population and organized environmentalists exist. However, none of the differences was surprising or seemed to be strongly based in ideological convictions. The differences, therefore, do not call into question the importance of the anomalies.

Within the environmental movement, however, we are a bit surprised to learn that members of new noncore organizations are more active than members of the other three types of environmental organization. The only exception is for conventional political behavior, for which there is no difference. The new noncore organizational type consists of three very different organizations: NOAH—for animal rights, Women–Environment–Development, and The Environmental Home Guard. Each of these organizations promotes a specific cause, or niche, of environmentalism (viz. animal liberation, ecofeminism, and responsible consumerism, respectively). These more narrowly defined goals might lead the members of these organizations to a broader environmental activity because increased focus can translate into greater commitment.

From these analyses, however, we are unable to conclude that members of core organizations are more active than members of noncore organizations, which was one of the expectations with which we started. When it comes to environmental behaviors, members of new core organizations are surprisingly passive compared to the activity of the members of new noncore organizations. Together with results from the previous chapter, this does not indicate that the core of the Norwegian environmental movement in any way presents itself as a radical alternative. A substantial difference is necessary to present a movement as an alternative to the existing order. Further, the results do not indicate that members of core organizations differ fundamentally from members of noncore organizations, nor do they differ fundamentally from the general population. All in all, we are unable to identify a strong and distinct pattern of beliefs behind the environmental movement. Why is this so?

The reasons for this, we believe, lies in the anomalies of the Norwegian environmentalism. The state-friendly society and the orientation toward local communities have moved the whole environmental movement, not the least being the core organizations, closer to the state and the government and, in turn, closer to the general population. Because moving closer to the state has had the effect of tempering environmental organizations, it is appropriate to return to a discussion of the anomalies of Norway's unique environmentalism.

Part III:
The Local Community Perspective in the State-Friendly Society

Chapter 7
The Local Community Perspective

Introduction

In Norway there is a tendency within environmentalism and nature protection to consider the human society, the local society, together with nature. The protection concerns humans at work, in close relations to wild nature. This is an extreme non-urban background, which is very difficult for foreigners to understand. This has led to controversies with Greenpeace and others because of sealing and whaling. Our whaling shocks them. For us this is a question concerning the protection of the values of north Norwegian coastal communities (Kvaløy Setreng, 1996, p. 110, our translation).

Few Norwegians share the popular view found in other countries that whales are part of a charismatic megafauna whose intelligence is second only to that of man. To most Norwegians, whaling is simply a way of harvesting from the ocean. In addition, it provides fishermen from small coastal communities with an additional source of income, which increases their community's chances of survival in the often inhospitable artic climate. Greenpeace tried to establish a local branch in Norway in the 1980s and 1990s, without much success (see Chapter 3). Greenpeace's animal rights policy fell on rocky places because few Norwegians understood why animals outside of man's interests should be attributed a privileged position.

The above citation from Sigmund Kvaløy Setreng, a leading environmental theorist in Norway, nicely illustrates the theoretical position that we label the local community perspective. Key aspects of this perspective are the intimate relationship between man and nature, that humans can survive using nature as a means of sustenance, and that small local communities, being the bearers of both rugged nature and Norwegian culture, are able to survive economically. We believe that the local community perspective can account for the weak position of animal rights in Norwegian environmentalism. Its roots can be found in the attempts of nation-building and the democratization process of the late 19th century. It is deeply interlocked with the development of Norway becoming a state-friendly society, with strong support given to the value of local government. This perspective is an intimate part of Norwegian political culture and a defining part of Norwegian environmentalism. To overlook this crucial dimension due to lack of contextual knowledge, as we think the Dryzek study does, might amount to a shortcoming. The absence of this dimension also means that state–society relations become too narrowly understood. Leaving out this important dimension leads to a lack of understanding of the specific character of Norwegian environmentalism and the role of civil society in general.

Center–Periphery and the View on Nature

Nation-states are geographical entities that link central and peripheral areas in a common political system of governance. The center is usually found around the capital, in which the most important political, cultural, economic, and financial institutions are also located. What is referred to as the periphery often moves beyond the smaller cities and larger towns and includes rural or nonurbanized areas located at some distance from the center. The more rural and the greater the distance from the center, the greater the periphery. The nation-state therefore demonstrates a more or less clear hierarchical structure, with central bodies and organizations at the top of the political/bureaucratic food chain of governance and with regional and local peripheral bodies subject to central control. Nation-state democracy has developed in a process by which the people are integrated into the governing of the nation by being included at the various levels in the hierarchical system (i.e., central, regional, and local). The democratic organization of the nation-state is thus the stable hierarchical order: Popular demands and interests are communicated bottom-up, whereas binding decisions are communicated top-down. In most South and Central European countries, the justification for this hierarchical order is as follows: The center represents the modern and progressive; the periphery represents the more backward and primitive. For this reason, the survival of the periphery must be ensured through its subjection to the political, cultural, and economic leadership of the center. Only in this way can the periphery be brought up to the developmental level of the center.

Norwegians, however, have traditionally held a slightly different view of the relationship between center and periphery (see also Chapter 2). True, Norwegians have also imagined that the farther away you get from the Oslo area, the further back in time you go. However, instead of going back to a primitive culture with no tomorrow, the voyage to the periphery has been interpreted as a journey back to the future: to a place where Norwegians found what is original and genuinely Norwegian (for instance, the idea of a glorious Viking past). One has to remember that for approximately 400 years, Norway was subject to Danish rule. In 1814, Denmark, fighting on the losing side during the Napoleon wars, ceded Norway to Sweden, and a personal union with Sweden was declared. This union lasted for more than 90 years. In 1905, Norway gained full independence as a constitutional and hereditary monarchy.

However, 500 years of foreign rule left its mark: The urban merchant and administrative centers were thought to have been contaminated by outside influence. The proper basis for a new national identity and culture was, therefore, believed to be hidden in the dimly lit rural peripheries—the parts of the country where the tentacles of alien supremacy had made little contact. This means that for Norwegians, the periphery was both primitive and modern at the same time in the sense that a reconstructed and synthesized version of periphery backwardness was catapulted to the apex of Norwegian nationhood. Consequently, preservation of the periphery has been seen as the proper defense of the nation.[113] The idea that the periphery is the cultural cradle of the nation has been manifested in, among other things, Norway's regional and agricultural policy, the temporal migration of Norwegians during vacations and weekends to their cabins in the mountains and along the fjords, Norwegian skepticism of the European Union, antiurbanism, and the wide acceptance of whaling and sealing.

The Norwegian hierarchical order has, as an outgrowth of this view, maintained three key features. First, the time frame has been one of historical continuity; that which is particularly Norwegian is found in the idea of the periphery's popular traditionalism rather than the center's avant-garde, elitist culture. Second, the ideological legitimacy of the periphery has led to political and economic power being spread out relatively evenly. In Norway, the area around the capital has not held the dominant political, cultural, and economic position that it has in many other European countries. Third, in a mountainous country where the center is looked at with askance and where human dwellings are few and far between, the nurturing of intermediaries like voluntary organizations and local government has been crucial for maintaining political unity. All this has had direct consequences for the structure of civil society and for the organization of state–municipal relations.

The period during which Norway was subject to Danish rule (1397 to 1814) is still tongue-in-cheek referred to as the night of 400 years. The year 1814, with the nascent country's constitution in place, symbolically marks the start of a national awakening. However, with the end of Danish rule, there was not much upon which one could build a national culture and identity. Fortunately, Norway rediscovered its heroic past of the Viking Age in which Norwegians (admittedly along with its Danish and Swedish brethren) ruled large parts of

---

[113] When Prime Minister Gro Harlem Brundtland in the run-up to the 1994 referendum on Norwegian membership in the European Union invited prominent members of the European Union to tour the northern Norwegian peripheries in order to demonstrate the need for these areas receiving transfers from the EU's regional fund, the strategy backfired. The EU dignitaries quickly became convinced that Norwegian peripheries were in no need of EU transfers or EU subsidies. In a European regional comparison, the standard of living in these areas were way above what could be observed in other regions. Norway was doomed to be a net contributor to the EU financial systems.

Northern Europe, and where Viking ships reached even further.[114] The Viking Age is still considered with awe and holds a dominant position in Norwegian mythology and history.

Being a people that, in its national myths, overslept the Renaissance makes it hard to construct a past of intellectual, cultural and civilizational progress. However, two significant survivors, or comparative advantages, of the 400-year night, one cultural and one natural, merged into a phenomenon of paramount importance. The first was the cultural one. For whatever reason, during their rule, Danish kings, compared to contemporary financial suppressors elsewhere in Europe, undertaxed the free Norwegian farmers to a degree that made them emerge buoyant when the new country gained independence. One cannot speak of an affluent class of independent farmers. Toil and moil on small farms was the daily routine. However, in the absence of a brokering nobility, free and independent farmers provided the country with a profound and almost unalterable egalitarian social foundation. In many ways, the nation had the history, culture, and geography on its side, and even under foreign rule, we can talk about a conscious awareness of Norway as a nation from the Middle Ages. Second, the natural advantage became evident when the young nation started to compare itself with the land of its former ruler. It lifted up its eyes to the hills and wondered from where help could come. There it was. The key characteristic of Norwegian natural landscape was mountains and fjords. Mountains reach for the sky. They contain the glaciers during summer, and they permit waterfalls to cascade down its walls and into the fjords below. In contrast to the flat Danish topography, mountains had a nationalistic potential that can not be overestimated.

Between the mountains of the land and the long surf-tormented shoreline of a vast ocean, the free and independent farmer, who toiled on meager soil, crystallized the image in which culture and nature coalesce. He comes in many disguises. He is the lonely man in a small boat in the open sea. Henrik Ibsen captures him in the poem *Terje Vigen* (1862). It takes place during the British blockade of Norway during the Napoleon wars. A man defies the blockade and rows his boat to Denmark to buy grain for his starving family, only later to be caught and imprisoned by the blockaders. He is also the frontier farmer who sets out to clear new land. Knut Hamsun captures him in *The Growth of the Soil* (1917).[115]

Nina Witoszek, a Polish historian working in Norway, argues, when discussing the role of nature in Norwegian knowledge and identity in the 19th century, that Norwegian patriotism at this time "... celebrated Nature as a source of national

---

[114] Norwegians argue that Leiv Erikson, who preempted Columbus' discovery of America by about 500 years, was a Norwegian, whereas Icelanders argue he was one of them. Following in the Erikson tradition of exploration, Fridtjof Nansen, Roald Amundsen, and Otto Sverdrup, among others, also contributed to nation-building and cultural consciousness raising by exploring the artic areas (see, e.g., Bomann-Larsen, 1995; Huntford, 1993, 2002).

[115] *The Growth of the Soil* (*Markens grøde*) earned Hamsun the Nobel Prize for Literature in 1920.

identity ..." and that "[s]o strong was the equation between nature and national-ity that in the 'politically correct' images of Norwegianness of the time, there was little room for an urban imagery." "Urban culture, associated with extraterritorial (i.e., Danish) clergy, bureaucracy and townsfolk, was alien to the folk spirit. It was nature, not culture, that was national" (Witoszek, 1997, pp. 214–215). Put simply, nature becomes culture. This is a lesser romantic view of nature than, for example, the German preservationist tradition. Rather, it is an emphasis put on the prominence of local areas. This tradition fostered the antiurban sentiment among Norwegians.[116]

In a sense, one can interpret Knut Hamsun's first book *Hunger* (1890/1899) as an antiurban manifest. Of peasant origin, Hamsun spent most of his childhood in the remote islands in Lofoten, in northern Norway. The opening line of the book has become a classic: "It was the period in which I was hungry and roamed Christiania [pre-1925 Oslo], this peculiar city who no one leaves without a mark." Having his articles turned down by every publisher, lack of food, and with an incipient insanity, the main character finally abandons the town and leaves in the hope of a better future elsewhere.

Humans and nature coalesce in the Norwegian context. The amalgamation emphasizes the positive aspects of local communities where the free farmer resides (see Tranvik & Selle, 2005; Østerud, 1978). It emphasizes the positive aspects of small local communities in contradistinction to an urban life. The special relationship to nature and its links to national history are also noted by David Rothenberg, another foreigner, who observes that Norway's "entire history is interwoven with the land. Norwegian national identity is nothing without nature" (Rothenberg, 1995, p. 201).[117] Witoszek observes that "[f]rom the organic perspective of dwellers in the land, their natural surroundings were less a romantic landscape and more a 'task-scape' ... , a dynamic man-in-nature gestalt imbued with action." And she adds that it "is not Nature seen through romantic lenses, but through pragmatic lenses, a nature that will deliver as long as we heed her and know exactly the horizon of limits to our interventions" (Witoszek, 1997, pp. 220 and 222).

This organic way of looking at the relationship between humans and nature is not based on romanticism, but on the way small local communities can survive in close relationship with nature. The local community perspective is based

---

[116] In 1993, one of the authors was asked to assist some Balkan refugees when they settled in Bergen after having to flee from their war-ridden country. To introduce them to their new cultural setting, they were invited to participate in the traditional Sunday walk in nature, which in the west coast city of Bergen area means a walk in a mountainous terrain. Not only were they surprised to learn that so many people were out for Sunday walks (it was, after all, a pretty day), but they were most surprised to learn that the people did not walk leisurely. People paced away as if it was a competition. Completely unprepared for such cultural practice, their experience and wonder with the Norwegian "ut på tur" ("go for a hike") fueled their conversations for weeks.

[117] This, in Rothenberg's opinion, provides one of the reasons for the Norwegian whaling.

on an understanding of humans and nature that is neither purely anthropocentric nor purely ecocentric (see Chapter 5). Although human beings can be regarded as being in the center, it is more of an equalization of humans and nature that can be labeled either "ecohumanism" (Witoszek, 1997) or "humans-in-nature." Nature is neither sacred nor divine. It is a place where people make a way of living and where they harvest and survive. Nature must not be exploited, and one must not extract resources beyond one's carrying capacity. The local community perspective of humans in nature is an irreplaceable context of modern Norwegian environmentalism. This is the more interesting because most environmental organizations are so centralized (some working only at the national level) and with a predominance of members living in urban areas (see Chapter 4).

The Exclusion of Animal Rights from Environmentalism

Norway experienced urbanization along with, but not to the same degree as, other industrial nations. Despite the country's urban development, a local community perspective was fueled by an ideology of populism and translated into the politically correct "regional policy."[118] In the Norwegian context, regional policy is a normative term that by default skews political debates on welfare and redistribution to the benefit of the periphery and leaves the burden of proof on any urban-tainted position. The regional policy stance, to a large extent, ensures that the standard of living in the peripheries will not drop too far below that of the urban areas, that people in the districts are offered decent employment opportunities, and that improved infrastructure (e.g., wider roads, bridges, tunnels, power supplies) facilitates communications with the anterior. Understanding local community and advocating regional policy improve the legitimacy and standing of political actors.[119] One should not forget that Norway is less urbanized than many other Western countries (even if a strong centralization process is now in the making). One of the consequences of the extensive regional policy is that Norway, a country of approximately 4.5 million people, still consists of more than 430 municipalities. Many of the municipalities are very small, consisting of a couple of thousand inhabitants, sometimes even less.

---

[118] The term "regional policy" is a somewhat inaccurate and loose translation of the Norwegian term "distriktspolitikk." By contrast, the term "urban policy" was nonexistent in Norwegian politics until the mid-1990s.

[119] Despite its urban base, some environmental organizations occupy themselves with the concerns of the districts. For instance, Nature and Youth advocates the upholding of small-scale farming and primary schools in the districts. Especially the closing of local post offices created heated debates between local communities and the central post authority. On the other hand, NOAH—for animal rights and Greenpeace Norway have been less preoccupied with local orientation (see Chapter 3).

Add to this political system the tradition of a very strong local government and we understand more fully how the system and tradition of local communities are politically entrenched.

We argue that the failure to graft the internationally successful environmental organization Greenpeace in Norway stems primarily from it being at odds with the local community perspective (as well as the state-friendly dimension; see Chapter 8). Internationally, the number of members of the organization was approximately 2.7 million in 1995 (Greenpeace, 1995; also see Chapter 3 of this volume). In Norway, it never accomplished a four-digit membership base. As a consequence, the organization never became part of the important networks of individuals that have developed through the generally extensive communications between state agencies and environmental organizations (see Bortne et al., 2001; Strømsnes, 2001; as well as Chapter 8 of this volume). Greenpeace Norway and its parent organization failed to understand or could not adapt to the fact that whaling and sealing have a long tradition in the survival of small communities, for whom harvesting from the fauna is a valuable source of income and employment. Not only the Norwegian public but also the environmental organizations and their members support a rational harvest of these animals. For instance, in our survey, about 80% of the general Norwegian population said that to prohibit the Norwegian seal and whale hunting was either "not very important" or "not important at all." In a 1993 national public opinion poll, two in three of the Norwegian population said that the relaunch of the scientific hunt for minke whales—the Norwegian in-your-face position toward the international whaling commission as well as the international community—was a correct government decision.[120]

It is also illustrative of Norwegian sentiments on this issue that the then leader of Nature and Youth (and later chairman of The Norwegian Society for the Conservation of Nature), Lars Haltbrekken, stated that "as long as there are enough whales, we can harvest them in the same way as we harvest elk and reindeer" (Haltbrekken, 1996, p. 159, our translation). Haltbrekken also willingly admitted that Nature and Youth had supported sealing at the beginning of the 1980s even though the toll for killing seals was so controversial. From an international perspective, support for the seal hunt is a somewhat surprising position for an organization that claims to be part of the deep ecological movement (Haltbrekken, 1996, p. 157). From a local community perspective, we do not think this position should raise any eyebrows.

Greenpeace's standing in public opinion in the early 1990s ranged from the conventional to the highly controversial. In a national public opinion poll in May 1991 on what the Norwegian population first and foremost associated with Greenpeace, 48% said environmental actions and environmentalism, 37%

---

[120] Source: Public Opinion Archives at Norwegian Social Science Data Services, June 1993.

reported actions against Norwegian whaling and sealing, and 3% answered that Greenpeace was "a terrorist organization."[121] Small wonder then that Greenpeace Norway had to close its Norwegian office in 1998.[122]

The fate of Greenpeace Norway has sharpened our understanding of the content and implications of the local community perspective. Specifically, the local community perspective, in which culture and nature coalesce, *excludes* animal rights from Norwegian environmentalism. In Norway, you can be a true environmentalist, profess environmental concern, and still support whaling and sealing. In other countries, by contrast, environmentalism to a greater extent *integrates* animal rights even if that does not always mean that there is much cooperation between animal rights groups and the rest of the environmental movement (Rootes, 2003). The exclusion of animal rights from Norwegian environmentalism, thereby contributing to the uniqueness of this brand of environmentalism, needs a second look. Here, we will use the animal rights issue to make inroads into the local community perspective (see also Chapter 9).

Two aspects of animal rights are important and relevant.[123] First, there is a general consensus in Norway that one should not impose harm on animals. Animal pain and suffering should be minimized as much as possible. However, such a view does not have to be associated with environmentalism at all. As long as the exercise of animal rights does not take place at the expense of local communities, the quality of life and well-being for domesticated as well as wild animals remain uncontested. Animal neglect and mistreatment is legitimately punished by Norwegian authorities. Animal husbandry is strongly regulated and controlled.

Second, if the exercise of animal rights conflicts with local interests in Norway, the default outcome is that animals must yield. Two types of problem illustrate a conflict between the needs of local communities and the natural needs and rights of animals. One problem is whether inhabitants of villages in northern Norway and some other local communities in the south should be permitted to kill whales for their meat. Basically, if animal needs and welfare conflict with human interests, the latter is favored. However, there is a vast difference

---

[121] Public Opinion Archives at Norwegian Social Science Data Services.

[122] In May 1998, the Finnish, Norwegian, and Swedish branches of Greenpeace merged. Their new headquarters is located in Stockholm. In 1999, they were joined by the Danish branch. Thus, the Norwegian branch of the organization has been dissolved, even though Greenpeace still holds a manned office in Norway. The Nordic Greenpeace organization has approximately 76,000 members in Sweden, 10,000 in Denmark, 1000 in Finland, and a couple of hundred in Norway (personal interviews; see also Greenpeace Norden, 2000, as well as Chapter 3 of this volume). At the end of 2005, these figures have increased so that the overall membership numbers in these countries are at approximately 105,000 members.

[123] A third aspect of animal rights can be discerned. It concerns implementations of animal rights in which adherents are willing to "liberate" animals from industrial farming by letting them loose, to resort to violence, or to damage persons and properties of butchers, slaughterhouses, and industrial farms. There have been very few such incidents in Norway. In the Norwegian context, such extreme behavior is quickly delegitimized due to the consensual mechanisms of political culture.

between small-scale harvesting on which small communities depend for their survival and large-scale harvesting by corporate industry and businesses, to which local communities are unimportant. More broadly, conventional hunting has a strong tradition in Norway. It attracts broad participation and has broad legitimacy. To the hunters and their families, the catch adds significantly to their household.[124] In this rather pragmatic and instrumental culture, it becomes irrational not to take advantage of resources that are readily available at almost no cost.

Another problem is whether wolves and bears should be allowed to stray in the Norwegian wilderness and prey on grazing sheep and cattle. This is a different context from the one discussed above and also different from the context in which whaling and sealing take place. It is more of a head-on conflict situation concerning whether predators should be able to take over the land where farmers fight for survival. This constellation is one that is more difficult for the government to regulate. This conflict might connect to other types of idea about nature and social organization. It should, therefore, be of no great surprise to find people supporting whaling and seal hunting while siding with the predators in their conflicts with the farming community. However, because of the way the local community perspective works, it is not a very common position to take.

The Norwegian Center Party with strong support in many rural areas, for instance, insists that wolves do not belong to the Norwegian fauna.[125] Although the size of the Norwegian wolf stock at the turn of the century was approximately 20–40 animals (Miljøstatus, 1999), the agrarian Center Party holds that wolves should be extinct or, at least, fenced because they cause severe problems for the farmers' livestocks. Strongly echoing the local community perspective, the party argues that the predators must yield if there is a conflict of interests between them and the domestic animals of local communities.

In addition to wolf and bear, the Norwegian predator fauna also includes lynx and wolverine. In a 1997 survey in Norway on attitudes toward predators, bears were the most popular animal. However, between 25% and 35% of the general population favored a reduction or the complete removal of bears, lynx, and wolverines from nature. Between 35% and 46% wanted the species to be kept at its present size, whereas between 26% and 40% wanted the stocks to be increased (Knutsen et al., 1998; see also Chapter 5 of this volume).[126] It is remarkable that in a country where there is a vast wilderness, close to one in three persons would

---

[124] In 1999, 350,000 hunters were registered in Norway. This is more than 10% of the adult population. On average 180,000 persons hunt every year, the vast majority (96%) being men between 25 and 45 years old. In a "normal year," the number of hunted animals is approximately 500,000 grouses, 40,000 elks, 23,000 deer, and 12,000 reindeer (personal interview with Espen Farstad, information officer in the Norwegian Association of Hunters and Anglers).

[125] In historical and biological terms, wolves in fact belong to the Norwegian fauna.

[126] Wolves were not included in the study, but, here, the negative sentiments are even stronger.

like to see a reduction in the stocks of predators. This is so even for people not having any of them in their local environment. Especially sheep farmers, on whose livestock predators also feed, were fiercely supportive of the reduction of predators. In another study, which showed a positive correlation between attitude toward predators and the NEP scale, sheep farmers endorsed ecological beliefs less than any other group (Kaltenborn, Bjerke, & Strumse, 1998).

In Norway, NOAH—for animal rights (NOAH) and partly Greenpeace face an uphill battle because their ideology runs counter to the local community perspective generally held by the inhabitants of the country. On the other hand, however, not even an organization like NOAH is in this inclusive polity treated as an outcast by the government. Although the organization of course tries to resist being co-opted by the state, NOAH willingly cooperates with the government on policies in which animal rights are addressed.[127] Even if NOAH to a larger extent than any of the other organizations is a political subgroup, it nevertheless turns to the government to improve environmental conditions. Conflicting views do not mean exclusion from the policy process. Neither does inclusion always result in moderation, contrary to what the Dryzek study (2003) holds.

Animal Rights Attitudes: Two Legs Good

In the subsequent analyses, we address attitudes toward animal rights that are so important in the concept of environmentalism in other countries. Due to its special position on these issues, we remove the members of NOAH from our typology and treat them as a category separate from the other four organizational types. This procedure prevents us from conflating organizational cause and specific environmental attitudes. Consequently, we expect members of NOAH to be most sympathetic to animal rights even when such rights come into conflict with the interests of local farming and livestock raising. One of the interesting aspects will be the degree to which the 11 environmental organizations in the revised typology deviate from the general population. If the local community perspective is important and broad, as we have argued, we expect differences to be quite small.[128]

It is important to distinguish between animal protection and animal rights. First, we asked the respondents whether they think animal protection is part of environmentalism. Except for NOAH, roughly two in three respondents agree strongly that this is the case (see Table 7.1, row A). On this topic, members are more inclusive than what we know from the policies of their organizations and from leadership interviews (Strømsnes, 2001). There are only small but statistically significant differences between the general population and organized environmentalists.

---

[127] For instance, NOAH takes part in governmental hearings (Bortne et al., 2001).

[128] Greenpeace is in this setting also deviant, but we have included the organization in the analyses because the organization is more than an animal rights organization.

For members of NOAH, 95% agree that protection of animals is part of environmentalism. This result shows the degree to which NOAH deviates from other environmental organizations. However, this is of course of no big surprise because animal protection defines the organization itself.

If animal rights are an integrated part of Norwegian environmentalism, one should expect environmentalists other than members of NOAH to include animal rights more substantially than the general population. However, they are hardly distinguishable from the general population. On the other hand, we are surprised to find that a majority in all groups claims that the protection of animals is a part of environmentalism in general. It looks like there exists a certain lack of symmetry between what the organizations as such express and what their members express. We believe this has primarily to do with the phrasing of the question and that *protection* of animals mainly taps the first aspect of animal rights in which absence of harm and pain is the criterion. Here, one should generally expect strong support. It is at the second level where animal rights conflict with local communities that we expected differences to kick in because

TABLE 7.1. Animal rights issues.

	GP	ONC	OC	NNC	NC	NOAH
A. Protection of animals is part of environmentalism (percent "strongly agree")	63	69	71	60	68	95
B. Protection that can prevent suffer and pain (percent "very important")						
Banning biotechnological "improvement" of animals [a]	63	74	72	76	73	90
Banning laboratory animals in the cosmetics industry	54	52	59	56	64	99
During extermination, limit suffering of insects/vermins [a, b]	48	40	38	48	42	66
Banning the use of animals in sports and entertainment [a]	22	17	21	20	25	70
C. Protection that can conflict with the survival of local communities (percent "very important")						
Protection of wild animals	54	68	84	66	80	91
Restricting industrial farming	30	58	49	55	50	86
Fighting all hunting not aimed at stock maintenance	29	28	30	31	35	71
Working against the fur trade	14	14	25	15	29	94
Banning Norwegian whaling	6	4	8	5	23	63
Banning Norwegian sealing	6	4	9	4	17	64

Note: NOAH is analysed separately from the organizational typology. N_{GP} = 945–1011, N_{OE} = 1962–2032.

[a] $p > 0.05$ between types of organization.

[b] $p > 0.05$ between the general population and organized environmentalists.

this is the area in which the local community perspective is most distinctly activated.

Therefore, we have sorted a total of 10 questions into 2 categories according to how we expect these questions to measure animal protection. The first category includes items that relate to the prevention of pain and suffering (see Table 7.1, row B). The second category includes items that can come into conflict with the survival of local communities (row C). This division follows our distinction of animal rights more closely and goes to the core of the local community perspective. The following items concern the animal rights aspect of protecting animals from suffering and pain; these items do not challenge the survivability of local communities: biotechnological improvement, the use of laboratory animals, painful extermination of vermin, and the use of animals for entertainment. The remaining items might tap more directly into the conflict between animal rights and survival of local communities. Protection of wild animals and a restriction on industrial farming will make it more difficult for farmers to make a living in some parts of the country. Hunting (especially of elk and deer) still makes a valuable additional source of income in many areas. The issues of fur trade, sealing, and whaling contribute specifically to the economy in some local communities and, therefore, to a greater extent signify the local community perspective.

The analyses show a clear difference between level of agreement with animal rights relating to prevention of suffering and pain (see Table 7.1, row B) and level of agreement with animal rights when in conflict with local communities (see row C). With the exception of the item "banning animals in sports and entertainment," there is among all groups a high agreement with animal rights as to prevent suffering.[129] When animal rights are framed within a local community perspective, the level of agreement drops for all groups. NOAH respondents' agreement on all issues loom high above the other groups, as expected. On all questions (except limiting suffering of insects and other vermin during extermination), organized environmentalists, more than the general public, show slightly higher agreement with animal rights issues. The item on protection of wild animals attracts surprisingly high agreement and more so in our survey as compared to other surveys mentioned in this chapter (see also Chapter 3). On the other hand, this question has not been explicitly contextualized as a potential conflict with domesticated animals. By contrast, agreement with a ban on fur trade, whaling, and sealing are very small. Restricting such activities would hamper the economy in local communities.

---

[129] To a Norwegian respondent, animals in sports and entertainment include circuses and horse racing. Because they appear more exotic and are found in other countries, cock fights, fox hunts, and bullfights, for instance, are in this context most likely not associated with sports and entertainment. To the best of our knowledge, no one in Norway has suggested copying the legal Spanish bullfight format and replacing the bull with a whale and within the confines of giant pools reenact old-fashioned whaling by permitting brave whalers in small boats to slowly but surely defeat the giant mammal. Alas, there is no tradition for barbaric animal shows in Norway.

None of the issues that we have classified as an indication of the potential conflict between animal rights and local communities in any way impinge upon life in urban areas. Because analyses in Chapter 4 confirmed organized environmentalists' urban roots, this result suggests that urban environmentalists are not at all unconcerned with local communities. This illustrates nicely the local community perspective within Norwegian environmentalism. In general, the lack of predominant agreement with animal protection that can conflict with the survival of local communities confirms the weak inclusion of animal rights as something more than animal welfare in the Norwegian brand of environmentalism. At least it is not part of any deeper ideology that might prevail over local interest if a conflict should surface. Within the field of organized environmentalism, animal rights as a deeper ideological force is a marginal cause. For the average citizen, images of childhood farm life are vivid and the rhetoric of local communities is still legitimate. In this mountainous country, the general animal rights cause, and the challenges to NOAH especially, more or less seem to be a continuous uphill task.

To simplify the remaining analyses, we have employed one single scale of animal rights and animal protection attitudes. The scale ranges from 0 to 1. Low scores are associated with disagreement with animal rights. High scores are associated with agreement with animal rights.[130] On the animal rights scale, the general population has a mean of 0.58, the organized environmentalists have a mean of 0.64, and NOAH has a mean of 0.92. Again, we see that organized environmentalists lie closer to the general population than to NOAH's extreme position. Controlled for the organizational typology, members of new core organizations have a mean of 0.67. In old core organizations, it is 0.64, in new noncore organizations (NOAH excluded), it is 0.62, and in old noncore organizations, it is 0.61. Thus, members of core organizations evaluate animal rights as somewhat more important than members of noncore organizations. However, in general, the differences are small.

Within the local community perspective, we bring two important variables to bear on animal rights attitudes: residence and distance to farm. First, for people in rural areas, there should be less tolerance for animals whose behavior and presence might conflict with the local populations' way of life. Due to the rural populations' residence and/or occupation, they should adhere more to the local

---

[130] We cast all 10 animal rights and protection items into principal component analyses in order to determine the degree of dimensionality. The analyses for both organized environmentalists (including NOAH) and the general population yielded one moderately strong first unrotated factor to which all variables contributed significantly. Eigenvalues and Cronbach's alphas for organized environmentalists and general population were 3.9 and 0.82, and 3.4 and 0.77, respectively. An additive scale was thus constructed for both groups. In subsequent analyses, NOAH is still treated individually. The construction of a single scale does not in any way violate our distinction of animal rights as related to both suffering and local community perspective. The distinction relates to different thresholds of support. The scale relates to the pattern of responses across all 10 items.

community perspective. For people in urban areas, animals in general or predators specifically might not impinge on their ways of life. Due to their residence and occupation, they should adhere less to the local community perspective. We therefore expect a moderate and positive relationship between animal rights attitudes and urban residence.

The analysis shows that such a relationship cannot be confirmed (see Figure 7.1). There is no support that urban residents have a higher concern for animals than do those who live in rural areas. For the general population, pro-animal sentiments are associated with smaller city residence only. Urbanity in general does not explain positive attitudes toward animal rights. This result shows that residence does not affect the local community perspective. There is no urban effect on the local community perspective. This result strongly illustrates how the local community perspective is generalized and deeply embedded in the way of thinking of those who live in this country. We believe this observation is a representation of something much deeper than the not-in-my-neighborhood principle.

Second, a variable that might combine physical and affective factors is one that we have labeled "distance to farm." It is measured by asking what generation in the respondent's family is the closest to include someone who grew up on a farm. We expect that respondents for whom there are several generations since anyone in their family lived at a farm are more concerned with animal rights than respondents who either live on a farm or whose parents lived on a farm. People who are inexperienced with farm life see the instrumental value of domesticated animals different from those who have observed, or even participates or participated, in the daily routines of farm life.

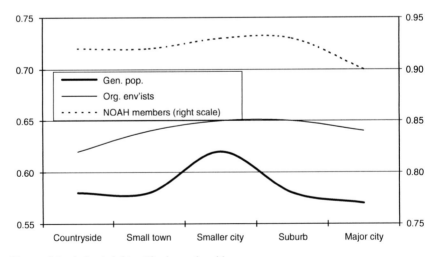

FIGURE 7.1. Animal rights attitudes and residence.
$N_{GP} = 933$, $N_{OE} = 1797$, NOAH = 157.
Note: NOAH (dotted line) on right Y-axis. Curves have been smoothed.

If we classify as being close to farm those who either live on a farm, have moved from a farm, or have parents who grew up on a farm, the distribution shows that 61% of the general population are close to farms. The figures for organized environmentalists and NOAH are 54% and 46%, respectively. First, these figures indicate that Norway has a recent agrarian past.[131] Second, the figures tentatively indicate that groups in greatest support of animal rights are also those with the longest generational distance to farms. This supports our hypothesis.

If we break our animal rights scale down to distance to farm, an interesting pattern emerges (see Figure 7.2). Because members of NOAH already show a very high agreement with animal rights (right scale), generational distance to farm only has a small marginal effect. For the general population, agreement with animal rights are mostly unaffected by distance to farm. For the main bulk of organized environmentalists, however, distance to farm has a significant positive effect. The longer the distance to farm, the greater is the agreement with animal rights (i.e., the weaker the effect of the local community principle).

Why is this relationship found for organized environmentalists only, not for the general population? Except for NOAH, it is only the *combination* of environmental membership and distance to farm that brings about increased animal rights

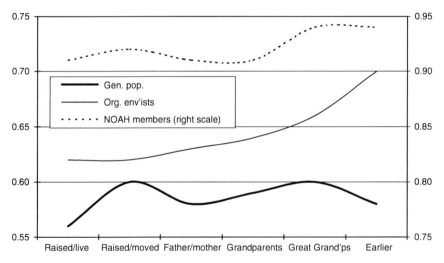

FIGURE 7.2. Animal rights attitudes and distance to farm life.
$N_{GP} = 920$, $N_{OE} = 1794$, NOAH = 156.
Note: NOAH (dotted line) on right Y-axis. Curves have been smoothed.

---

[131] If we also include "grandparents on farms" as being close to farms, these numbers increase to 82% for the general population, 81% for organized environmentalists, and 77% for NOAH.

sentiments. This indicates that membership, which is a result of general environmental concern, is a necessary condition for high animal concern. However, we cannot decide whether this is best explained by organizational socialization or "self-selection" (that members had these attitudes before joining the organization). Somewhat paradoxically, these data suggest that a general increase in environmental concern and subsequent environmental membership will diminish the local community perspective, which makes Norwegian environmentalism unique. If this is the case, organized environmentalists might develop in a direction different from the general population and, in turn, challenge the present political consensus of regional policies. As a consequence, one might encounter increased political conflict. In the longer run, however, we might see a similar development within the population at large. Perhaps there is a certain time lag only, because distance to farm will simply increase among the population. What we observe might be that we are in the midst of a more profound weakening of the local community perspective (see Chapter 9).

We also explored four variables that might correlate with the animal rights scale. These variables can cast more light on the variation we observe.[132] First, we recoded religious beliefs into a dummy variable where "I do not believe in God" and "I am ignorant about God's existence" were coded 0. All other options, which in one way or another contain some sort of belief in God, were coded 1 (see Figure 5.7). For both the general population and organized environmentalists, those with religious beliefs evaluated animal rights as more important than atheists and agnostics (see Chapter 5 for a discussion on religion). NOAH members' attitudes toward animals were uninfluenced by their religious score.

Second, radicalism, measured by the left–right self-placement scale, correlates with animal rights for both the general population and organized environmentalists. A preference for left politics therefore coalesces with higher concern for animal rights. Consequently, people who place themselves at the political right evaluate animal rights as less important. Again, there is no significant relationship for members of NOAH.

Our analyses also show that there is a positive relationship between ecocentrism, measured by the 7-point self-placement scale (see Chapter 5) and animal rights sentiments for both the general population and organized environmentalists. For members of NOAH, again the relationship is insignificant because their agreement with the animal rights scale is already very high.

Fourth, we related attitudes toward abortion with the animal rights scale. This variable is not necessarily far-fetched. The logic behind this approach is that people who for various reasons have restrictive views on the termination of pregnancies might be more reserved in intervening in the lives of animals and nature in general. Abortion is measured by a composite scale where low

---

[132] Here, we should keep in mind that these four variables in no way tap the whole local community perspective because animal rights is just one, albeit crucial, dimension measuring a kind of opposition to local community attitudes.

scores are associated with restrictive views on abortion and high scores are associated with liberal views on abortion.[133] The analyses show that there are significant negative relationships between the abortion scale and the animal rights scale for the general population and members of NOAH. A restrictive view on abortion therefore coalesces with a positive attitude toward animals, and vice versa. For organized environmentalists other than members of NOAH, there was no relationship.

Finally, we cast all variables addressed in this section into a multivariate regression analysis to test for their significant effects when controlled for the other variables. We also add sex and age as controls to check for the importance of gender and being young, as their importance is so central in much of the literature on environmentalism.[134] For the general population, the analysis shows that variables that had significant effects on animal rights were a restrictive view on abortion, ecocentrism, radicalism, and being a woman. For the organized environmentalists other than members of NOAH, significant effects were distance to farm, ecocentrism, radicalism, age, and being a woman. For NOAH, the only significant effect was a restrictive view on abortion. It is interesting that views on abortion is the only variable that can affect the already high level of NOAH members' position on animal rights. This suggests that for members of NOAH, protection of animals and the unborn child are strongly related.[135]

Taken together, the important message is that the distance to farm variable, after multivariate control, remains positive and significant for organized environmentalists only. For members of NOAH, no other correlate than views on abortion is able to display variance in their strong commitment to animal rights. This strong commitment turns members of NOAH into strong outliers in the environmental movement, furthermore, it gives insight into why they are not included in the more general understanding of Norwegian environmentalism.

Conclusion

In Norwegian history, nature has been linked to how inhabitants fight for survival in local communities. Nature has strongly influenced what is national and

---

[133] The composite abortion scale consists of six questions (i.e., the child's sex is not the desired one; the woman does not want the child; the family has a very low income and cannot afford more children; the woman's life or health is at risk; there is considerable risk of the child suffering a serious illness/handicap; there is a risk of the child suffering a less serious illness/handicap; the woman has been a victim of incest or rape). The one-factor principal component solution for the organized environmentalists (NOAH included) and the general population yielded an eigenvalue of 3.28 and 2.99, and a Cronbach's alpha of 0.82 and 0.79, respectively.

[134] Residence is not included here because it did not correlate with animal rights in the first place.

[135] Note that we found 75% of the members in NOAH to be women (see Chapter 4).

influenced the inhabitants' self-understanding and political culture. The more rugged the survival in nature, the more genuine the definition of Norwegians. The relationship between nature and humans has become special when nature is a means of survival. For instance, people who depend on nature for a living do not have the burden of proof when human interests come into conflict with conservation interests. It is a paradox that predators like the wolf and bear, when preying on marginal and unfenced sheep and cattle, cause such a heated discussion in a country with such a vast wilderness. However, in Norway, a local perspective entails that urban interests, at least up to the present, play a role secondary to those of the local communities. However, it also entails that animal rights is very weak within Norwegian environmentalism.

We have used attitudes toward animal rights as a test of the local community perspective. Unsurprisingly, organized environmentalists evaluate animal rights as somewhat more important than the general public. They are also distinguishable in that order when we discern animal rights as protection from suffering and pain as opposed to animal rights as a threat to local communities. When animal rights issues conflict with the survival of local communities, they are evaluated as less important than issues that protect animals from suffering and pain. This illustrates the local community perspective in Norwegian environmentalism in which the differences between the organized environmentalists and the general population are very small. We attribute their similarity to the local community perspective in which Norwegian environmentalism, including the Ministry of the Environment, and consequently, most environmental organizations, have excluded animal rights from the definition of environmentalism. That is so even if the general understanding of environmentalism in Norway is extremely broad and includes much that are not commonly understood as environmentalism (see Chapter 2 of this volume, as well as Bortne et al., 2002). The organization NOAH, whose members consistently evaluate animal rights as being very important, thus becomes an outlier of the environmental movement in the same way as is the case with Greenpeace.

For organized environmentalists, the important factor contributing to an increased concern for animal rights is generational distance to farm. The longer the time since an ancestor left a farm, the stronger the animal rights sentiments. Over time, therefore, one might anticipate that organized environmentalists will move closer to the animal rights position of NOAH, thereby distancing themselves from the general public on this view. Consequently, one might expect to see a greater conflict between the general public and organized environmentalists on animal rights issue. Even so, as we have discussed earlier, in an era in which urbanization continues and Norway gains more experience with agribusiness and as the distance to farm necessarily increases as time goes by, it might be that the general population will follow suit so that no new and deep conflict between environmentalist and the population at large occurs. However, because a decline of the local community perspective in the minds of people does not necessarily mean that these people become animal rights activists, the consequences of the decline of the local community perspective is too early to call.

Whatever these changes might bring, they are bound to bring about a profound change in the Norwegian political culture. Consequently, one cannot rule out that Norwegian environmentalism might lose some of its distinctiveness regarding the absence of animal rights from its brand of environmentalism. The type of change we are anticipating here will not be unconnected to what happens to the other anomaly.

Chapter 8
The State-Friendly Society

Introduction

In the Scandinavian countries, the states not only play a dominant role, but they are also open and inclusive and heavily influenced by societal groups and open to citizen requests. Inversely, the citizens of these countries are rather friendly and open-minded to the state in which they live. In Norway, the relationship between the state and environmental organizations (or any type of voluntary organization) cannot be understood without what has been coined as the "state-friendly society" thesis (see Chapter 2 of this volume; Kuhnle & Selle, 1990, 1992b; Tranvik & Selle, 2005). Although many organizations in these countries receive a large part of their income directly from the government, the organizations' share of government income in the Scandinavian countries is smaller than what similar organizations receive in most EU countries (Sivesind et al., 2002). Within this Scandinavian system, government and authorities are—within limits, of course—open-minded to criticism of their environmental policies. It is not the position of being critical that excludes an organization from being able to cooperate closely with governmental bodies; it is more the fact that the organization and its members are not understood to fall within the overall definition of environmentalism itself.

To what extent, then, do Norwegian environmental organizations and their members consider the state an opponent or an ally in the environmental battle? We argue that the prevailing state-friendliness tempers the environmentalists and pulls environmental organizations closer to the state and to the political decision-making. Whatever the organization's original and deeply based opposition to government, this proximity makes opposition not only more difficult but also less desirable if the organization is looking for political influence. This open political system—being much more open than what the Dryzek study (2003) gives credit—provides both autonomy and influence to organizations concerning their decision-making and implementation of their environmental agenda. Generally speaking, we are dealing with a political system that in a comparative perspective definitely is a "thick" rather than a "thin" democracy (see Chapter 2) and that includes a specific type of state–civil society relations. However, as we will discuss more fully in Chapter 9, this political system might now gradually become "thinner."[136]

In this chapter, we address the second anomaly of Norwegian environmentalism: state friendliness.[137] We examine how organized environmentalists evaluate the relationships between environmental organizations and the state. We also address the problem of why there is no green party in this ostensible green polity. Finally, because state-friendliness, environmentalism, and perception of democracy are intricately bound together, we analyze the environmentalists' assessments of the internal democracy of their organizations, the limits of democracy in general, and the ways in which both might impede environmental progress.

The Financial Dilemma of Organized Environmentalism

Due to its nonprofit orientation, a voluntary environmental organization most often depends on a source of income other than what it technically produces. If we stick to the sector triangle of state, market, and civil society, an environmental organization can receive money from any of them. An organization's sources of income from the civil society will most likely stem from membership dues or members making monetary donations. For many organizations, buying and selling goods are also important.[138] This source will make the organization more independent of financial support from the state or market actors. However, due to Norwegian environmental organizations' rather low membership rates, an organization cannot rely on its members alone. In addition, some of the new organizations have deliberately turned away from the tradition of trying to maximize its membership base. The member marginalization strategy will make the civil society as a sole source of income improbable for most of the recently established and nondemocratic

---

[136] In Tranvik and Selle (2003), we argue that although the state is as strong as ever and has not really been weakened by sundry globalization processes, it is the representative democracy that is losing ground. The general trend in this process is that public bureaucracies and specialized professions increase their power and influence at the cost of elected bodies (see also Tranvik & Selle, 2005; Østerud et al., 2003).

[137] In order to assess whether there exists a relationship between organizational structure and attitudes toward state, democracy, and other organizations, we constructed a complementary organizational typology for the analyses in this chapter. The complementary typology reflects the internal organizational structure in a different way. From the original typology, we replaced the age dimension with a democracy dimension. The consequence was that World Wide Fund for Nature moved from "old core" to "nondemocratic core" due to its lack of internal democracy (see discussion in Chapter 2). In all relevant analyses in this chapter, both typologies were used. The analyses showed that only small differences between the two typologies could be discerned. Therefore, we only report the results from our original typology as presented in Chapter 2.

[138] For a detailed analysis of the financial sides of the Norwegian voluntary sector, see Wollebæk and colleagues (2000) and Sivesind and colleagues (2002).

environmental organizations. In this development, we also see that new "check-book" or "protest business" organizations might generate money from nonmembers in civil society (Jordan & Maloney, 1997; see also Chapter 6).[139] Thus, for sources of income to secure its operations, an organization must also turn to the state, to the market actors, or both. Of particular interest here are the consequences that an organization must face when turning to one of these sources.

Normally, there are two types of state financial support for Norwegian voluntary organizations: *project grants* and *basic grants*. If an organization receives project grants from the state, the state earmarks money for specific projects. This commits the organization to a project that the state has scrutinized and approved. Subsequently, the organization can actively develop and implement the project. Project grants lead to a deeper integration between state and organization. Basic grants simply provide the organization with a general financial support that can be used at the organization's own discretion, such as the daily operations of the organization. However, basic grants also impose a certain "political correctness" on the part of the organizations, such as internal democracy and a guaranteed minimum level of organizational activity. However, the development we now observe in the direction of increased use of project support under the auspices of the new neo-liberal state might put a new type of pressure on organizational autonomy. This fact makes the insights from the Dryzek perspective (2003) increasingly relevant. This development is, to a large extent, a consequence of the introduction of New Public Management principles in the public sector. Here, we see a move away from governmental bodies that generally have placed great trust and autonomy in voluntary organizations and toward a new type of pressure on organizational autonomy in which governmental bodies increase their control on how the organizations spend their money and whether the money is spent in an efficient way (Tranvik & Selle, 2003, 2005). In adapting to these changes and to fulfill the new government-imposed requirements, organizations tend to become increasingly centralized and professionalized.

Some organizations, most notably The Bellona Foundation, have claimed that receiving project grants make them less dependent of the state than do basic grants. We do not share this view. Due to the inflow of state money as project money, once the organization finds itself within the reach of the state tentacles, it could prove difficult to maintain a credible and legitimate opposition to the state. Project grants can, therefore, jeopardize the independence of environmental organizations. Moreover, as the organization adjusts its expenses to its income, it is hard for an environmental organization to present itself as being in radical opposition because project grants can convert the organizations into semistate agencies and, by extension, implementers of public policy. In these developments, we see the contours of a new type of control regime.

---

[139] Among the 12 environmental organizations, only Greenpeace Norway operated without financial support from the state or the business community. The Norwegian branch of Greenpeace received money from its mother organization Greenpeace International.

Financial support from private companies presents itself as an alternative to state grants. However, market grants can switch organizational dependency from the state to market actors. Sources of income from several market actors, therefore, seem better than only one market benefactor, as dependency will decrease as the number of contributors increases. For instance, Norwegian environmental organizations allow various sponsors to buy advertisements in the organizations' newsletters and magazines. Also, some environmental organizations, most notably Green Warriors of Norway, now accept money from private-sector companies and act as environmental consultancies in return (Strømsnes, 2001).[140]

If the choice of funding is freely available to an organization, too great a reliance on market actors over state funding can be risky in that the former can withhold their funding were an organization to be involved in activities or issues unfavorable or discrediting to these actors. Even worse, a market actor can veto an organization's project were it to compromise, or in any way be in conflict with, the goals of that actor. Critics of The Bellona Foundation, which is the Norwegian environmental organization that to the largest extent has cooperated with market actors, have said that this organization has no credibility left because it serves as a consultant company for the business sector (Søgård, 1997; Tjernshaugen, 1999). To these critics, there seems to be a clear trade-off between an organization's cooperation with market actors and its environmental credibility. In response, The Bellona Foundation claims that it does not operate as a consultancy because financial funding by market actors is not linked to certain projects. In view of the neo-liberal transformation of the public sector and the government demand for increased audit and control, environmental organizations are guaranteed not to be able to avoid this strategic dilemma over funding and survival (Strømsnes, 2001).

Between Fragmentation and Unity

However, the question of whether certain dilemmas are perennial to organizations also depends on whether the types of organization are the same and whether the voluntary sector considers itself as part of a larger organizational entity. In Norway, voluntary organizations work closely together with the state. If the state was monolithic, as implied by the Dryzek study, the perspective from the state should therefore be that such organizations constitute one cooperating sector. However, because governmental bodies are increasingly segmented, even fragmented, with little coordination across policy fields (e.g., Østerud et al., 2003), what governmental bodies often prefer is cooperation between organizations

---

[140] Some voices in the civil society literature argue that the modern types of organization are leaving their members behind and organizing themselves increasingly in such a way as not to become dependent of their members either financially or in terms of member activity. In this perspective, it is not the members who are leaving the organizations, but the other way around. See Papakostas (2004) for an interesting discussion on this topic.

within different policy fields. This preference actually fits with the current situation within the voluntary sector. According to a recent study, the voluntary sector in Norway has, for a long time, not considered itself as being one larger unit (Sivesind et al., 2002). Rather, the various organizations within the voluntary sector consider themselves divided into different fields according to what type of service or activity they produce (e.g., health, culture, or environmentalism). In contrast to developments in Britain and the United States, where organizations develop identities as a moral power outside of and in opposition to the market and the government, associations in Norway as well as in the other Nordic countries only to a small degree share a common identity (Wollebæk et al., 2000).[141]

The Norwegian version of state-friendliness thereby generated a proximity between the state and the voluntary sector. However, the institutional links between the organizations and the state differ across policy fields. The environmental field has been one in which strong integration has occurred. However, this proximity was not found among organizations across different fields. This observation undermines the chimeric notion of the voluntary sector making up a "people's movement." It is therefore pertinent to commence our analyses of the state-friendly society by addressing the question of whether members of environmental organizations consider the advent of new movements such as the peace movement, the environmental movement, and the gay/lesbian rights movement as a continuation of the voluntary movement or as a general change into more distinct movements and movement cultures.

The questionnaire offered a 7-point scale that was recoded into a 0 to 1 scale. Low scores indicate that the organized environmentalists think that the new movements are all part of one overall movement. High scores indicate that the organized environmentalists think that the new movements are clearly distinct from one another. The mean for all organized environmentalists is 0.55. This suggests that the new movements, at least to some extent, are perceived to be distinct from one another. Mean scores range between 0.50 for new noncore members and 0.58 for new core members. In other words, members of new core organizations are most particularistic, whereas members of new noncore organizations are most inclusive. This means that members of the single-issue organizations in the new noncore group see a larger coherence among these new organizations than do members of new core organizations. Perhaps these organizations need the other organizations to be able to see themselves in a wider context and that these organizations are more conscious in perceiving all types of new movement as being part of a larger whole. This is probably also one of the reasons why their members do participate so extensively in voluntary activity in general (see Chapter 6). To frame this in the language of the social capital literature, members of the core organizations tend relatively more toward "bonding" (stay with your own), whereas members of the noncore and single-issue organizations tend more toward "bridging" (more open to others) (Putnam, 2000; Wollebæk & Selle, 2002a).

---

[141] See also Seip (1984) and Lorentzen (1998).

Relationship with the Authorities

If the state-friendliness thesis is valid, one should observe that members of environmental organizations favor close relationships between the organizations and the political authorities and that voluntary organizations are important, even necessary, for a good society. First, the results in Table 8.1 strongly indicate that both the general population and organized environmentalists to a large extent favor close relationships between organizations and authorities. All in all, this points in the direction of symmetry between members' attitudes and organizational behavior. A close relationship is not something that governmental bodies force on organizations, as implied in the Dryzek perspective. Rather, cooperation is simply considered the natural way of doing things. The item "close cooperation with the authorities leads to the organization's loss of credibility and freedom to act" underscores this tendency because it receives little support. In addition, at least 9 in 10 respondents across all organizational types, including the general population, agree that voluntary organizations are a prerequisite to a good society. These numbers are consistently high. They strongly suggest that there is little if any state skepticism in Norway. However, this should not at all be understood as Norway having a population accepting anything that has a government label. What we argue is that it is the populations' overall cognitive horizon that is state-friendly. Within this system, of course, ideological differences as well as conflict over issues can still exist.

It is remarkable that among organized environmentalists, for instance, one is able to observe that at least three in four agree with an item that not only permits

TABLE 8.1. Relationship between organizations and the authorities; percent "agree".

	GP	ONC	OC	NNC	NC
Voluntary work and voluntary organizations are a prerequisite to a good society.[a]	90	95	96	95	94
The authorities should be provided with information on and control over environmental organizations' use of government funding.	87	75	71	76	66
It is best for society if the voluntary organizations work closely with the authorities.	83	80	80	80	69
The authorities must provide the necessary financial support to voluntary environmental organizations.	74	77	79	82	75
Close cooperation with the authorities leads to the organization's loss of credibility and freedom to act.	30	34	31	36	42

[a] No statistically significant differences between types of organization at the 0.05 level.

$N_{GP} = 817–932$, $N_{OE} = 1899–2012$.

but also accepts that "authorities should be provided with information on and control over environmental organization's use of government funding" and "it is best for society if the voluntary organizations work closely with the authorities." These results might be somewhat surprising in an international context. However, in a state-friendly society as the Norwegian one, close relations to the state is, for most organizations, the only way to survive both in terms of funding and legitimacy. It is not only a tactical strategy, it is the standard operating procedure. Similar patterns of such a preferred link between voluntary organizations and the state is also confirmed in other studies covering the voluntary sector as a whole (Wollebæk et al., 2000, p. 238).

We observe that members of new core organizations deviate the most from members in the other organizations. Members of new core organizations are somewhat more skeptical of the relationships between organizations and the state than members of any other organizational type.[142] This suggest a certain pattern between members and organization in that the action-oriented behavior is more frequently favored by members of these organizations (see Table 8.1) and their confrontations with the authorities (which we discussed in Chapter 3) might have a negative influence on the members' evaluation of the state. We believe that this pattern is part of a more general ideological change that is not only limited to the environmental field.[143]

Organizational Measures for Environmental Ends

Members of environmental organizations are at least occasionally expected to think about the importance of measures for their organization to succeed in promoting its policies. A measure can be defined as a resource that might enable one to reach certain ends. Organizational measures can be assessed differently according to what type of organization one is a member. One distinctive difference between measures is whether they are legal. A measure such as achieving media coverage is probably gaining ground because it is important to an increasing number of organizations and it seems to become more and more important to be visible in the media for organizations to survive. This measure might be specifically important to new and nondemocratically built organizations, which more than other organizations seem to need constant exposure to retain their support.

Organized environmentalists evaluate different organizational measures as consistently more important than do the general population (see Table 8.2). Between 40% and 49% of the general population consider all measures but illegal ones as very important. The figure for illegal actions is 6%. Among organized

---

[142] In Chapter 4, we observed that the members of new core organizations are least frequently employed in the public sector. This *might* also be linked to their more critical attitude toward the state.

[143] For further discussions along these lines, see Tranvik and Selle (2003) and Wollebæk (2000).

environmentalists, all measures except legal protest and illegal actions are considered very important by no less than 50% across all organizational types. With a view to the issues that involve interaction with governmental authorities, the results can be seen as another indication of organized environmentalists being state-friendly. Although sheer cooperation does not necessarily entail mutual confidence and trust, extensive cooperation over long periods of time would be difficult in the absence of such trust (Rothstein & Stolle, 2003).

Media coverage seems to be evaluated as equally important as contact with the authorities. Roughly two-thirds of all environmentalists state that achieving media coverage is very important. Although media coverage is primarily associated with new core organizations, members of old noncore organizations are only slightly behind the former on this issue. This figure shows a surprisingly similar assessment of the importance of media coverage. The fact that three of five members of old noncore organizations agree that media coverage is important for environmental organizations shows that environmentalists have become aware of the importance of getting the message across to the public. Perhaps media coverage to a certain extent is a compensation for the lack of strong organizational democracy. Even among the general population, media coverage is considered more important for an environmental organization than having contact with the authorities in various ways. Relatively few members of old noncore organizations find measures of legal protests and demonstrations and illegal actions very important. It is the new core organizations that are the most polarized on these measures.

Across the organizational measures for one's organization to succeed in promoting its policies, members of new core organizations deviate most consistently from members of the other organizational types. Members of new core types of organization emphasize the importance of "unconventional" activities, such as lobbying, demonstrations, and civil disobedience, more than others do. "Conventional" activities, such as taking part in implementing government environmental policy and

TABLE 8.2. Importance of measures for your organization to succeed in promoting its policies; percent "very important".

	GP	ONC	OC	NNC	NC
Achieving media coverage	49	59	67	70	72
Taking part in implementing government environmental policy	48	56	58	67	49
Having direct contact with the authorities in order to try to influence them (lobbying)	47	66	73	70	78
Taking part in government-appointed councils and committees	43	62	62	60	52
Taking part in legal protests and demonstrations	40	24	42	41	51
Taking part in illegal actions (civil disobedience)	6	6	14	12	27

Note: In the general population survey, "your organization" was replaced by "environmental organizations." All groups are statistically different at the 0.05 level.
N_{GP} = 962–968, N_{OE} = 1985–2018.

taking part in government-appointed councils and committees, is deemphasized. Even if the differences are not dramatic, again this pattern indicates that new core organizations represent a type of environmental activity different from both the general population and the other three organizational types. This is also to some extent expressed through the members' attitudes and behaviors (see Chapters 5 and 6). Furthermore, these observations also fit in with what we see in many other types of new voluntary organization. The new way of doing business is an important part of the ongoing transformation of the voluntary sector.

Most Important Area to Affect

From the issue of which measures should be taken to promote organizational policies, we also asked the respondents which areas they believe are most important for their organization to affect. In general, "public attitudes towards the issues the organizations works on" and "decision made by national government" were considered the two most important areas (see Table 8.3). Less priority was given to areas involving regional or local government, international forums, and business.[144]

However, priority patterns differ between types of organization. The members of new noncore organizations find public attitudes significantly more important than other members do (62%), whereas influence on decisions made by national authorities is considerably lower (17%). This reflects, in part, the environmentalism policy of consumerism of The Environmental Home Guard, which belongs in this type of organization. Members of old noncore organizations find decisions made by the national government more important to influence than do members of other types of organization (41%). This figure almost equals the priority given

TABLE 8.3. Most important area to affect (percent).

	ONC	OC	NNC	NC
Public attitudes toward the issues the organization works on	45	43	62	32
Decisions made in the national government	41	27	17	25
Decisions made by regional/local government	7	9	10	8
Decisions made in international forums	5	19	9	21
Decisions made by business	2	3	3	14
Total	100	101	101	100

Note: Questions were not posed to the general population.
Groups are statistically different at the 0.05 level. Percents sum to more than 100% due to rounding.
$N_{OE} = 1354$.

---

[144] These questions might not clarify the finer distinction between who should make the crucial decisions and who should implement them. This might explain why the local level is not coming out stronger.

to public attitudes (45%). Although the figures are generally low, new core members are significantly overrepresented in giving priority to "decisions made by business" (14%). Not only do new core environmentalists reveal some skepticism of the state (see earlier discussion), they also, more than other environmentalists, turn toward decisions made by business actors.

Decisions made by international forums are not given high priority by the respondents. However, there is a clear difference in the responses in that members of core organizations seem to be more oriented toward international forums than members of noncore organizations. Although this pattern suggests that core organizations have a more distinct international profile, one should expect these numbers to be significantly higher in all groups because, almost by definition, environmental problems are international in nature. The finding that environmental protest is mainly national in character is in accordance with what the Rootes study (2003) found for some EU countries. It is therefore not surprising that a lack of strong international orientation also describes the Norwegian voluntary sector in general.

The response pattern for the environmentalists observed in Table 8.3 seems to confirm that Norwegian environmental organizations mainly operate between the state and the local communities. The response pattern also illustrates the weak international orientation and priority among Norwegian environmentalists and, by extension, environmental organizations (see discussion in Chapter 7). Although both the priorities given to international forums and the business community are low as to the most important areas to affect, one can not rule out that this will change. New core organizations, whose members give higher priority to these areas than do members of other organizations (21% and 14%), might, due to their being in the center of the movement with effective organizations, suggest the direction in which the next generations of environmentalists might pursue. We find it a bit surprising, if not a little depressing, that with the global character of environmental problems and with all of the international negotiations going on in which Norwegian government play a rather important role, the environmental part of the Norwegian voluntary sector is still rather introvert.

Decision-making and Decentralization

An extension of the topic of which area is most important to affect is *who* should make the most important decisions on environmental issues and which measures are most decisive for the solution of environmental problems? As to the first issue, respondents show a fairly high rating for all agencies under consideration (see upper half of Table 8.4). This is unsurprising considering the nature of the issue at hand and the lack of constraints on the items (because respondents theoretically could agree on all). In light of the local community perspective discussed in Chapter 7, one should note that personal and local decision-making are given somewhat higher ratings than supranational decision-making. Approximately 75% of all environmentalists agree that delegating authority to local communities can solve environmental problems, albeit not necessarily the most fundamental ones. This result suggests a confidence in the decisiveness and importance of local communities in at least certain types of environmental affair.

The respondents also had to choose which single measure is the most decisive for the solution of environmental problems (see lower half of Table 8.4). Compared to the responses in the upper body of the table, the picture is now reversed. International cooperation is singled out as the most desired solution, especially among the general population. Organized environmentalists put approximately equal weight on the international solution and that individuals change their lifestyles. The local solution is given the lowest priority as to being the most decisive solution of environmental problems across all categories.

There are several reasons for the ostensible differences in the result between the two parts of Table 8.4. One reason is the distinction between how one perceives the world and how one wants the world to be. An individual might prefer local solutions, at the same time admitting that international solutions might be necessary once a single decisive solution has to be implemented. This is also reflected in the rating and ranking responses offered in the survey on these questions. In these answers, we see the problems that many respondents have in putting ideas and ideals into a coherent picture. In general, however, the response pattern we observe is also what can be expected when Norwegian environmentalism is understood by way of the local community perspective. When forced to opt for a decisive solution, people are sufficiently realistic to opt for an international solution. This does not at all imply a strong international orientation in people's daily thinking (see Chapter 5). However, when environmentalists are forced

TABLE 8.4. Evaluation of decision-making and decentralization; percent "agree".

	GP	ONC	OC	NNC	NC
Decisions on environmental affairs must be made by, or in close collaboration with, those affected.[a]	88	84	85	83	82
We must delegate authority in environmental affairs to local communities.[b]	78	78	73	80	72
We must delegate authority in environmental affairs to supranational bodies.	59	74	73	65	73
Most decisive for the solution of environmental problems?					
Extensive international cooperation	63	41	46	34	46
That individuals alter their lifestyles	24	43	42	55	37
That the national authorities give higher priority to environmental protection	8	13	10	8	15
That the local authorities give higher priority to environmental protection	5	3	3	3	2
Total	100	100	101	100	100

[a] No statistically significant differences between types of organization at the 0.05 level.
[b] No statistically significant differences between the general population and organized environmentalists at the 0.05 level.
Percents exceed 100% due to rounding.
$N_{GP} = 865–972$, $N_{OE} = 1829–1980$.

to make a choice on such difficult topics, it is reasonable to expect that realism might win. As in other areas of life, you cannot always get what you want.

Trust in Institutions

Trust, meaning someone or something in which confidence is placed, is "widely considered to be of fundamental importance in political explanation" (Eckstein, 1997, p. 32). It is a crucial variable within the new social capital literature (Putnam, 2000; Rothstein & Stolle, 2003; Stolle & Hooge, 2004). Trust in institutions depends on the socialized attitudes and worldviews of individuals, but probably also on the experience with the institutions as such.[145] Previous research has shown that Norwegian citizens' trust in government and private institutions are among the highest in Europe (Listhaug, 2005; Listhaug & Wiberg, 1995; Strømsnes, 2003). Due to the state-friendly society thesis, we assume that what organized environmentalists might lack in trust in international and economic institutions (due to their lack of international orientation and their market skepticism), they make up in trust in government institutions. Among a number of different institutions that can be trusted, we are particularly interested in the extent to which organized environmentalists and the general population trust four different institutions: first, the national political system, which primarily represents the state and which, in turn, buttresses state-friendliness; second, the political parties, which are the final national arbiters of environmental policies; third, the environmental authorities, which implement and audit the environmental policies; fourth, the voluntary organizations, which hold the potential as being watchdogs and correctors of the state. These items are italicized in Table 8.5.

The results are very interesting indeed. Organized environmentalists trust the national political system and the political parties *more* than do the general population. Organized environmentalists, who are supposed to represent organizations that are alternatives to and correctors of the state, trust the state to a larger extent than individuals who have not joined such organizations. That must obviously have something to do with the open and inclusive character of the state. These results are strong evidence of the existence of a state-friendly society. In addition, approximately 55% of both the general population and the organized environmentalists trust the environmental authorities. It would have been reasonable to expect that organized environmentalists would be more reluctant to trust authorities if they felt that the authorities were doing less than what is needed. Again, if there were to be skepticism of national implementers of environmentalism, the

---

[145] Whereas Rothstein and Stolle (2003), for instance, argue that generalized trust to a large extent is explained by personal experience with institutions, Uslander (2002) held that trust is foremost a result of childhood socialization and that later experience is less important. Most social capital researchers place themselves somewhere in between these two positions.

TABLE 8.5. Trust in institutions; percent "trust" (much/fair amount).

	GP	ONC	OC	NNC	NC
The Norwegian educational system	80	81	78	73	72
The Norwegian legal system (including courts)	77	87	86	78	79
Our national political system[a]	*77*	*84*	*83*	*80*	*79*
Voluntary organizations[a]	*76*	*92*	*95*	*95*	*95*
The United Nations (UN)	71	60	68	64	60
The Norwegian military[a]	71	53	49	47	50
Environmental organizations	62	na	na	na	na
The Norwegian state church	59	50	46	48	34
Norwegian environmental authorities[b]	*56*	*52*	*55*	*58*	*45*
The business community[a]	50	31	32	32	31
The European Union (EU)[a]	34	21	24	23	27
The press and mass media[b]	32	28	36	30	36
Norwegian political parties	*29*	*35*	*38*	*33*	*29*

[a] No statistically significant differences between types of organization at the 0.05 level.
[b] No statistically significant differences between the general population and organized environmentalists at the 0.05 level.
Note: Item wording: "How much confidence do you have in the following institutions?"
na: not asked.
$N_{GP} = 921\text{--}982$, $N_{OE} = 1911\text{--}2001$.

environmentalists would not take it out on the national environmental authorities. We are unable to decide whether the extent to which the high level of trust is strengthened by the direct experience with governmental bodies or if it is a matter of self-selection. Finally, environmentalists trust the voluntary organizations significantly more than the general population. Within voluntary organizations in Norway, there has been a feeble understanding of being part of a societal sector in its own right. Yet, members of environmental organization show strong trust in other voluntary organizations. We have also found in the voluntary sector at large a high degree of trust in the types of voluntary organization of which respondents are not a member (Wollebæk et al., 2000).

Among the general population, it is only the European Union, the mass media, and, somewhat surprisingly, political parties that have a level of trust less than 50%. The other institutions' degree of trust range from 50% to 80%. These figures ensure that a general level of trust is maintained across a number of institutions and that society does not at all seem to suffer from a lack of legitimacy.[146] Among organized environmentalists a number of interesting findings can be noted. In general, organized environmentalists trust various institutions to the same extent as the general population. If the Norwegian environmental movement was an alternative movement or a subculture, we should have expected a significantly lower level of institutional trust. A conspicuous exception to this is the

---

[146] These findings are supported by findings from another general survey on Norwegian citizenship in 2001 (Strømsnes, 2003).

level of trust in the business community, in which 50% of the general population reports trust, in contrast to only one in three among organized environmentalists. This result is probably largely explained by the leftist orientation within the movement (see Chapter 7).

These results on the potential scope and role of the voluntary sector tie in with what has been observed earlier (Wollebæk et al., 2000). Among the general population there is not much support for the idea of a voluntary sector not being tied to the state and the market, or a voluntary sector without state funding. Wollebæk and colleagues (2000) found that a clear majority of the population wants the voluntary organizations to cooperate closely with both state and market. However, this cooperation should not be done in such a way that these organizations take over new and important tasks at the expense of the public sector. Thus, the voluntary organizations cannot present themselves as a clear-cut alternative in opposition to the state. On the contrary, they still need the state as a source of money and, by extension, legitimacy. This entails that they are cognitively oriented toward the state. These observations, which might also apply to the other Scandinavian countries because the patterns we observe there are so similar to those we have found in Norway, force us to pursue the structural relationship between the state and organizations. In our opinion, these findings are crucial in order to clarify the relationship among different political actors and to fully understand the society in which the Norwegian environmental organizations operate. It is quite another setting than described by the Dryzek study. Within an environmental perspective, we find the electoral failure of the Norwegian green party to be particularly telling. We use the absence of a green party to provide insight into the dynamics of the system at large.

Accepting the Issue—Dismissing the Agent

The Norwegian multiparty system rests on a number of cross-cutting cleavages. This system is supported by an electoral system of proportional representation with fairly large constituency sizes. Therefore, the threshold for gaining parliamentary seats is fairly low.[147] The postwar period brought few but significant changes in the Norwegian case of the Scandinavian five-party system (Berglund & Lindstrøm, 1978; Lindstrøm, 1997). A Christian People's Party had already been founded in 1933, whereas a New Left Party and a Progress Party were formed in 1959 and 1973, respectively. A number of viable small and medium-sized parties

---

[147] For example, in the 1993 election, the Liberal Party received 3.6% of the votes, whereas the Red Electoral Alliance received 1.1%. Both parties entered parliament with one seat each. In the 1997 election, the Liberal Party received 4.5% of the votes and six seats because it passed the 4% national threshold, thereby qualifying for the allocation of additional seats. The new pro-whaling Coastal Party ("Kystpartiet") entered parliament in 2001 with one seat after having received only 0.4% of the national vote, but 6.2% of the vote in the county of Nordland.

are therefore present in this system, but without any of them being exclusively based on a single issue.

Although it was not until 1989 that the recently established Green Party took part in a national election in Norway, the prospects of such a party seemed good. During the 1989 election campaign, 37% of the voters reported that the environmental issue was the most important issue to them, and environmentalism ranked second in importance among all available issues. However, when the Green Party entered the 1989 general election, it received only 0.4% of the votes and gained no seats. In the 1993 election campaign, only 7% of the voters reported environmentalism to be important to them, thus providing it with a fifth place among contested issues (Aardal & Valen, 1995). In this election, the Green Party received only 0.1% of the votes. In the 1997 election and the 2001 election, 0.2% of the voters voted for the Green Party. In local elections, the Green Party has gained only a handful of representatives scattered across some city councils nationwide. Like Greenpeace, the Green Party never was able to get a foot in the door of the important and established political networks. Therefore, the Green Party has not been a part of the formal and informal structures that are so important in a state-friendly society (Selle, 1999). In sum, the Norwegian Green Party has not had any electoral success. By contrast, several other European countries, including neighboring Sweden, were able to foster more successful green parties (Båtstrand, 2005; Müller-Rommel, 1998; Richardson & Rootes, 1995; Rüdig, 2006; Wörlund, 2005).

The main Norwegian parties are sometimes ordered along a left–right dimension (see, e.g., Gilljam & Oscarsson, 1996; Grendstad, 2003b). Within this ordering, the main parties are: Red Electoral Alliance ("Rød Valgallianse," RV), the Socialist Left Party ("Sosialistisk Venstreparti," SV), the Labour Party ("Det norske Arbeiderparti," DNA), the Center Party ("Senterpartiet," SP, the former Agrarian Party), the Liberal Party ("Venstre," V), Christian People's Party ("Kristelig Folkeparti," KrF), the Conservative Party ("Høyre," H), and the Progress Party ("Fremskrittspartiet," FrP). However, why does the alleged green Norwegian polity lack a significant green party? Here we limit the discussion to highlight three factors only: the absence of a nuclear issue, the anchoring of green politics on the ideological left, and the openness of the political parties (see also Knutsen, 1997; Aardal, 1990).

First, all Norwegian parties favored nuclear energy in the 1960s. Even the nature conservation organizations approved nuclear energy because they considered it to be an important alternative that would save remaining waterfalls from being developed for the production of hydroelectric power (Knutsen, 1997). However, on Christmas Eve 1969, North Sea oil explorers reported that they had struck black gold. Politicians quickly learned that energy, and in turn money, could be supplied in abundance from the oil reserves in the North Sea. In addition, one also developed techniques so that hydroelectric power could be produced and used more efficiently. With the energy questions being fought over the rate of oil exploration and development of hydroelectric power, the removal of the nuclear issue from the political agenda deprived the Norwegian environmental movement of an issue around which they could have rallied support. By comparison, the nuclear issue became highly politicized in Sweden. In many countries, it has proved to be one of the most controversial issues along the environmental cleavage (Müller-Rommel, 1985; Rootes, 2003, see also Chapter 3 of this volume). The Norwegian Green Party is so insignificant that an important study on the electoral success on green parties in Europe simply excluded the Norwegian case (Müller-Rommel, 1998), whereas another study included an analysis of the members of the Norwegian green party (Grendstad & Ness, 2006). In Norway, we argue, the nuclear issue was put on the back burner. In turn, the politicians never came to challenge the environmental interests on this issue (Knutsen, 1997).

Second, during the first two postwar decades in Norway, there was no real opposition to the Labour Party's policies of reconstruction, industrialization, and economic growth.[148] In 1970, when the preservation-based environmental movement successfully drew attention to yet another majestic waterfall to be piped for the cause of hydroelectric power (i.e., "Mardøla-aksjonen"; see Chapter 3), the increasing environmental consciousness was subdued by the emerging major political issue of the decade: the question of whether Norway should join the European Economic Community (EEC). The outcome was to be decided by a referendum in 1972.

Environmentalism does not have intrinsic properties by which we could settle the issue of its ideological home. However, since the Norwegian environmental movement more strongly rejected capitalism than socialism, the EEC issue cemented legitimate environmental opposition at the ideological left of Norwegian politics. This political positioning was welcomed by left-wing populist movements that had merged the EEC issue with the political advocacy of the sparsely populated Norwegian districts. Simply put, this was the local community perspective at work (Brox, 1966; Knutsen, 1997; Aardal & Valen, 1995).

In the 1973 election, left-wing populism successfully organized itself into the Socialist Electoral Alliance (which included the Communist Party) and gained 11.2% of the votes. This alliance was dissolved at the next election. Thus, when green opposition emerged from consensual politics in early 1970s, the EEC issue effectively bounced it back onto the radical side of the left–right dimension.

In Chapter 5, we saw that organized environmentalists placed themselves considerably to the left on the left–right axis compared to the general population. This supports the "watermelon hypothesis" that environmentalists are green on the outside but, once the jacket is removed, the rest is all red (or at least pink). In general, this observation also corroborates Poguntke's claim that "New Politics groups are generally situated on the left of the political spectrum" (Poguntke, 1993, p. 11). The analyses in Chapter 5 also confirmed that primarily the Socialist Left Party, but also the Liberal Party and the Red Electoral Alliance were

---

[148] In the postwar period, the Labour Party was the dominating party intermittently serving as a single party government for close to 40 years.

overrepresented among organized environmentalists. The Labour Party was most strongly underrepresented among organized environmentalists, followed by the Conservative Party, the Christian People's Party, the Progress Party, and the Center Party. Hence, if we follow a conventional ordering of parties from left to right (which we outlined earlier), parties on the left are overrepresented and parties on the right are underrepresented in terms of organized environmentalists' party preference. However, the conspicuous exception is the Labour Party. This is no big surprise because historically the Labour Party for long has been the dominant party of economic growth, large-scale industrialization, and technological optimism.[149]

In Chapter 5, we also saw that 95% of all environmentalists were willing to place themselves on a left–right scale. This indicates at least that the scale is highly "recognized" in that respondents show a "willingness and ability to place oneself on the left-right scale" (Fuchs & Klingemann, 1989, p. 208). Here, we examine to what extent organized environmentalists assess the left–right scale as a good or bad representation of the political landscape.[150] Roughly 50% of organized environmentalists consider the left–right axis to be a bad representation of said landscape (see Figure 8.1). This proportion varies little across party preference. We are therefore somewhat surprised to learn that the voters of what election researchers consider to be the two Norwegian "green" parties [i.e., the Socialist Left Party and the Liberal Party (Aardal, 1990)] do not distinguish

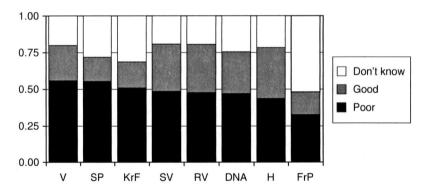

FIGURE 8.1. Assessment of the left–right scale by party preference. Organized environmentalists only.
$N = 1588$.

---

[149] See also Grendstad and Strømsnes (1996) and Grendstad and Ness (2006).
[150] In the subsequent analyses, we address the members of the environmental organizations as voters. Our aim here is to test whether or not there exists an electorate from which the Green Party can draw support. Thus, we have not used the organizational typology in this part.

themselves from voters of other parties with respect to whether this political dimension reflects the political landscape well.

We find that organized environmentalists with a preference for the Progress Party find the left–right dimension least favorable. However, environmental voters of the Progress Party also revealed the highest share of "don't knows." The parties are slightly differentiated by the percentage of those who find the dimension to be a good representation of the political landscape. This group surprisingly consists of adherents of the Socialist Left Party, Red Election Alliance, and the Conservative Party. In short, political party adherence does not separate organized environmentalists when one assesses the validity of the left–right dimension. This observation seems to convey a broad political consensus within the environmental movement.

The simple correspondence between the dimensions of left–right and environmental protection versus economic growth can be expanded to include an analysis of the degree to which the general population and organized environmentalists' party preferences correspond along said dimensions. This approach will inform us on the degree of difference between political parties along the two dimensions across the general population as well as the environmentalist population (see Figure 8.2). The upper right corner combines the political "right" with a preference for economic growth. The lower left corner combines political left and preference for environmental protection. Each line in Figure 8.2 represents party preference. It connects the general population and organized environmentalists (shown by

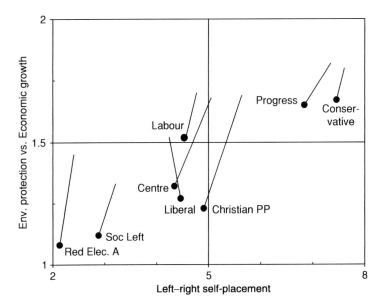

FIGURE 8.2. Old and new politics. Party preference and environmental membership. $N_{GP} = 777/810$, $N_{OE} = 1589/1609$.

a bullet). For instance, for the Christian People's Party, we see that the general population place itself further to the right and further progrowth than do the organized environmentalists who prefer this party.

Basically, the political parties fall along a positive-sloped diagonal indicating that there is a significant degree of correspondence between the left–right dimension and the dimension of economic growth versus environmental protection. Predictably, parties on the left are associated with environmental protection and parties on the right are associated with economic growth. In a sense, there are three party clusters along the left–right dimension (i.e., Red Election Alliance, Socialist Left versus Center, Liberal, Labor and Christian, versus. Progress and Conservatives) and two clusters along the environmental dimension (i.e., Progress, Conservative, and Labor versus the other parties). Without exception, we also observe that environmentalists are more in favor of protection than the general population within all parties. It makes a clear difference whether one is an organized environmentalist or not, even if we cannot gauge the effects of organizational socialization versus self-selection. Also,with the exception of the Liberal party, environmentalists are more radical than the general population at large in all parties.

Finally, the established political parties began early to incorporate conservationist and preservationist issues in their programs. In the beginning of the 1960s, the program of the Conservative Party had a larger emphasis on classical conservationist issues than any other party program in Norway (Bjørklund & Hellevik, 1988; Knutsen, 1997; Sejersted, 1984). However, in the period from 1969 on, the Liberal Party and the Socialist Left Party have alternated on devoting the most attention to environmental protection. Despite its verbal commitments to environmentalism, the Labour Party has been regarded as the party for economic growth *par excellence* (Knutsen, 1997). The Conservative Party never let go of its focus on the importance of economic growth. It trails the Labour Party closely with regard to the politics of economic growth. The main point here is the rapid reorientation of some of the political parties to adapt to changing political times. In a sense, the Liberal Party and the Socialist Left Party have occupied and preempted the political space that would otherwise most likely have been the turf of the Green Party. Furthermore, the lack of a nuclear issue around which to mobilize does not make party success easier for a green party.

Openness is not limited to political parties. In many ways, the government bureaucracy also is open and easy accessible. Ideas from environmental movements and other new social movements have been, and for the most part still are, quickly accepted and incorporated by political organizations, political authorities, as well as within nonenvironmental voluntary organizations. Social protests are sometimes transformed into reforms before the protestors are given time to organize politically into a social movement, not to speak of a successful political party (see Chapter 1 of this volume, as well as Gundelach, 1993; Tourain, 1987). This is exactly what we have seen in Norway. This process influences organization building, ideology building, and consequently, the recruitment of members.

At other times, political authorities might legitimize organizations' policies even if these seem to be at odds with the government's own policies.[151] The general message is that the openness of the political structures more or less dovetails the pervasive public attitude in which many legitimate interests can be regarded as a public concern either in terms of political problem-solving or receiving financial support.

The next analysis on party preference combines some of the above issues. Earlier, we saw that organized environmentalists and the general population did not differ significantly with regard to trust in various institutions (see Table 8.5). In fact, the negligible differences might indicate that the environmental movement gains little legitimacy when establishing itself as an alternative movement independent of the state. Here, we analyze the degree to which trust in institutions differs across preference for political parties. Such an analysis will enable us to observe whether trust in institutions differs between, on the one side, the Liberal Party and the Socialist Left Party, both of which are the polity's functional green parties, and, on the other hand, the other political parties.

Here, we are particularly interested in four institutions: the national political system, environmental authorities, political parties, and voluntary organizations. These represent the most important institutions of the state-friendly society. We combine the eight party preferences with our two populations of organized environmentalists and the general population. We thereby obtain 16 categories. By using discriminant analysis, we seek to establish which type of institutional trust separate the 16 categories (see Table 8.6). The first dimension, which separates the most among the 16 categories, is a general dimension of trust in national institutions: the environ-

TABLE 8.6. Trust in institutions; discriminant analysis.

	National institutions	Organizations and parties
Norwegian environmental authorities	0.84	−0.17
The national political system	0.69	0.19
Norwegian political parties	0.67	0.47
Voluntary organizations	0.03	0.81
Eigenvalue	0.14	0.07
Percent of variance	57.50	31.30

Note: Two discriminant functions requested only. Cell entries are discriminant loadings.
$N_{GP} = 703$, $N_{OE} = 1443$.

---

[151] Attac was founded in France on 3 June 1998. Attac Norway was founded on 31 May 2001. The organization participated in the at times violent demonstrations at the EU summit in Gothenburg, Sweden, 14–17 June 2001, where the president of the United States, George W. Bush, also was present. At that time, Norwegian Minister of the Environment Siri Bjerke, at a ceremony in Oslo on 14 June of the same year, delivered the so-called Sophie Prize to Attac International. In an official statement by the Ministry issued on 2 June, Bjerke quoted the principles of the prize and unambiguously stated that "the Prize shall go to individuals or organizations that in a pioneering or a particularly creative way, has pointed to alternatives to the present development. From what I have learned, ATTAC definitely fulfills these criteria" (Bjerke, 2001).

mental authorities, political system, and political parties. This dimension, then, combines both diffuse trust in the political system and specific trust in political bodies. The second dimension combines trust in specific organizational actors. One type of actor are the parties competing for political power and the other type are the organizations within the specific political field of voluntarism.

The distribution of the 16 groups reveals a distinct pattern (see Figure 8.3). Because most lines are positively sloped and organized environmentalists within each party are closer to the top right corner (identified with a bullet), organized environmentalists trust voluntary organizations, political parties, and national institutions *more* than the general population does. Again, we see no empirical support whatsoever for the claim that the Norwegian environmental movement is an alternative movement in opposition to the state.[152] The cognitive horizon of environmentalists is one that is aligned *with* the state, not away from the state.

Further, the distribution of parties shows that the Progress Party, conventionally found on the political right, and the Red Election Alliance, conventionally located on the political left, unite in a relative distrust of national institutions. In a sense, these wing parties are in opposition to the political system. In addition, adherents of the Progress Party among the general population have significantly

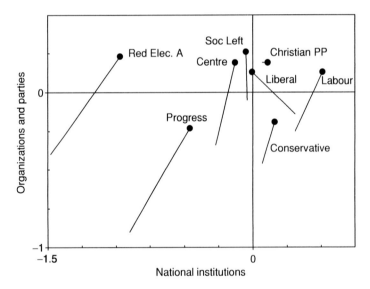

FIGURE 8.3. Trust in institutions; discriminant analysis.
$N_{GP} = 703$, $N_{OE} = 1443$.

---

[152] In fact, it should come as no surprise that organized environmentalists trust their own species. However, this trust is not at all tempered or compensated by distrust in political parties, some of whom evidently have co-opted the environmental issue at the cost of a potentially electorally competitive green party.

less trust in organizations and parties compared to all other groups. The other parties are clustered nicely together. Differences between organized environmentalists and the general population are conspicuously small. Environmentalists in the Labour Party trust national institutions the most. Environmentalists in the Liberal Party deviate from all other parties in that they show less trust in national institutions than do members of the general population who adhere to this party.

Conclusions from these analyses can be drawn at two levels. First, the absence of the nuclear issue as a catalyst, the anchoring of environmental issues on the political left (and a failure to reject the left–right distinction), and the openness of the political parties (and other types of voluntary organization) give strong clues as to the absence of a significant green party in an alleged green polity. The Socialist Left Party and the Liberal Party are the functional green parties in Norway.[153] There does not seem to exist any "political space" left for a pure environmentalist party.[154] Second, organized environmentalists trust the state—in these analyses represented, among others, by the environmental authorities and the national political system—*more* than does the general population. Elsewhere, this result would most likely have been interpreted as an error of coding or error of analysis. However, in Norway, the result is surprising only if one fails to understand the state-friendly society of which organized environmentalists are a part. We do not—as the Dryzek study (2003)—understand this high level of trust primarily as an effect of co-optation that makes environmentalists gradually more moderate as they cooperate with governmental bodies. Because of the overall political culture that the organizations operate within, Norwegian environmentalists are moderate from the start even though they are found to be more radical in certain periods than others (see Chapter 3). The variation in radicalism is more related to broad and external features in the general environment. The observation that organized environmentalist surpass the general public in trusting the state, we believe, solidly buttresses our state-friendly-society hypothesis. If trust in the state had been lower among environmentalists than among the general population, the prospects for a green party would have been better. Consequently, the state-friendly society must wither if a green party is to succeed.[155]

Organized Environmentalism and Democracy

One of the dilemmas between environmentalism and democracy is often posed as one of *efficiency*. The question is whether democracy is a suitable tool or an

---

[153] In September 1995, after the Swedish election to the European Parliament, the former leader of the Socialist Left Party in Norway, Erik Solheim, referred to the Swedish Left Party ("Vänsterpartiet") and the Swedish Environmental Green Party ("Miljöpartiet de Gröna") as "our two sister parties in Sweden" (Knappskog, 1995).
[154] The term "political space," as used by Skocpol (1992) and Sklar (1993), can be defined as "the fit between political institutions and group capacities" (Skocpol, 1992, p. 54). See also Berven and Selle (2001).
[155] Futher comparisions with Sweden will clarify the Norwegian case here.

inefficient tool for solving environmental problems. To some environmentalists, the environmental crisis is so pressing that solutions should be reached instantly and not through democratic procedures involving protracted hearings, deliberations, and negotiations. There is simply not enough time to let the environmental crisis work its way through the democratic process. Democracy is not suitable when it comes to convincing people about the extent of the environmental problems or for implementing the required and necessary measures (Lafferty & Meadowcroft, 1996). Democracy, therefore, is not only "a method for those who are patient" (Gleditsch & Sverdrup, 2002), it is also a method for those who fail to see the severity of the environmental crisis.

Related to this point of view is another of democracy's dilemma: democracy's intrinsic *value* compared to other societal values. In what way can one advocate that democracy, which was created under quite different conditions and adapted to other preindustrial and preglobalized settings than those we have today, is the right institution to deal with radical ecopolitical action? Is it correct to assume that democracy, which some hold might be partly to blame for the environmental problems we presently face, will also be able to solve the very same problems (Wyller, 1999)? Faced with an irreversible environmental catastrophe, it might simply be untenable that democracy can be accorded a superior value in its own right. How long can democracy permit itself to aggravate the environmental crisis without undermining its own foundations? If we stick to democracy as a procedure to uphold social life, perhaps there is a risk that, in the end, we could lose both. For those who interpret the environmental development as heralding a deep and irreversible conflict, there is a value hierarchy where life itself supersedes democracy. If this is the case, the latter must yield. Although green ideology and environmentalism do not take democracy for granted, it is, on the other hand, not unconditionally discarded (see, e.g., Dobson, 1990; Dobson & Lucardie, 1993; Doherty & de Geus, 1996; Lafferty & Meadowcroft, 1996; Mathews, 1996; Paehlke, 1996).

A third democratic dilemma is whether all those who will be *affected* by today's democratic decisions on the solutions to environmental problems have been properly represented in the decision-making. Our form of democracy demands that the interests of our fellow citizens must be represented. Today's decision-making procedures, however, do not give a voice to all those potentially affected by today's decisions. However, several people contend that in the ideal green society, consideration must be given to the interests of citizens in other countries, to future generations, and even to nature itself.[156] Much literature has been devoted to such intergenerational justice (eg, Brundtland, 1987; Malnes, 1990; Lem, 1994; Lafferty & Langhelle, 1995).

---

[156] Luc Ferry has been highly critical of the idea that animals and nature should be given rights. Nature is not an agent of its own (a tree cannot sue a human; another human must sue on behalf of the tree), and it is almost impossible to make clear distinctions among different forms of life (1996, p. 170). Rhetorically, Ferry asks whether the HIV virus too should be given rights.

Earlier in our analyses, we found that voluntary organizations in general are trusted by a large majority of all respondents. The analyses also showed that voluntary organizations are regarded as a prerequisite for a good society. This provides a basis for a strong civil society in which social capital is built.[157] When we link discussions on democracy to voluntary organizations, three aspects can be discerned. First, participation in organizations is educational in that organizational activities might function as "schools in democracy." Members are educated about the rules and procedures of organizations and democracy. This provides them with "political capital" (Gundelach & Torpe, 1997; Selle & Øymyr, 1995). Second, the more internal democracy in an organization, the more the organization at large will represent the will of its members. Third, an organization's external democratic function (or role in the overall political system) is met when the organization mobilizes new groups into politics and when it sustains the interests of its members. In so doing, the organization contributes to bolster the plurality of the civil society.

In the next section, we address two aspects of perceptions of democracy related to organized environmentalism in Norway.[158] First, we study how the environmentalists assess the internal democracy of their organization. Second, we assess democracy external to the organizations and assess democracy as a value, democracy as efficiency, as well as the degree to which democracy is fair. We emphasize especially whether organized environmentalists have perceptions of democracy different from those of the general population. If so, can this difference be linked to the organizational types to which the environmentalists belong? We also study whether we are able to ascertain the profound change in the understanding of democracy that emerges with the new generation of organizations from the mid-1980s?

Democracy Inside Organizations

A key question concerns whether an organization should maintain and improve internal democratic values and procedures vis-à-vis its members or whether democratic values should be minimized so as to employ the organization's resources for the sole benefit of the environment. The litmus test is often whether members in the organization are considered a *resource* or a *problem*. In the former case, it is only through a large body of members that an organization can achieve its

---

[157] The large Johns Hopkins Comparative Nonprofit Sector Project has developed a Global Civil Society Index in an attempt to grasp the strength of civil society organizations in different countries. The index consists of three dimensions: "Capacity," "Sustainability," and "Impact." Of the close to 40 countries included in this index, Norway comes second after the Netherlands (Salamon & Sokolowski, 2004).

[158] In our subsequent discussions on democracy and environmentalism, we take the survey questions on democracy as our points of departure. We do not aim to engage in any general discussions on democracy.

goals. This is the classical organizational structure that has been dominant since before the turn of the 20th century. In the latter case, members are a "problem" because attention to them deprives the organization of time and resources from its environmental goal. By extension, because democracy seldom reaches consensus but is a procedure to resolve conflicts, minimization of internal democracy might also reduce the scopes of organizational conflicts.

During the 1970s, democracy and efficiency were, to a large extent, regarded as two sides of the same coin. In the 1980s, democracy and efficiency became more of a dilemma necessitating a choice between the two. The Bellona Foundation, which we have classified as a new core organization, is a case in point. One of the prominent arguments in favor of its founding in 1986 was the desire to create an organization that used less resources on internal bureaucracy and more resources on real environmental efforts. Its founders wanted "greater freedom of action than they could get in a democratic organization" (Persen & Ranum, 1997, p. 93, our translation), claming that using time to recruit members was simply "not worth it" (Nilsen, 1996, p. 194). These positions are very good examples of the shift from democracy to efficiency that coincided with the growth of the new types of voluntary organization from the mid-1980s on. We expect this organizational distinction to be reflected in the data.[159]

The analysis shows that members of new organizations more than members of old organizations agree that democracy impedes environmental tasks (see Table 8.7, Panel A). Although the differences in the data are not great, the results corroborate the expected difference between old and new types of organization.[160] However, this difference is not observed when members are to assess organizational conflicts. Four in five members of new core organizations acknowledged that conflicts are a natural part of an organization's life. This is perhaps the reason why they are members of organizations in which the role of members and democracy are minimized.

Approximately four in five respondents agree that the leadership can act without consulting the members and that the leadership is good at dealing with conflicts within the organization (see Table 8.7, Panel B). These figures initially indicate that leadership has high legitimacy across the organizational types. Members of core organizations agree the most that there is too great a distance between leaders and followers in the organization. However, only members of new core organizations find that leadership provides inadequate information and acts too much on its own. Members of new core organizations especially seem to find themselves in a dilemma where the low level of internal democracy is obtained at the expense of an autocratic leadership. Overall, however, leaders of voluntary organizations have a high legitimacy in Norway.

---

[159] Items on internal democracy were asked to members of the organizations only.

[160] In these analyses, we are unable to control for self-selection (e.g., that environmentalists who deemphasize democracy opt for nondemocratic organizations) or that "world time" in the 1980s [to use an expression from Skocpol (1979)] is conducive to a new kind of organization and a new kind of member.

TABLE 8.7. Evaluation of internal democracy; percent "agree/never".

	ONC	OC	NNC	NC
Percent "Agree"				
Panel A				
Running an environmental organization democratically takes too much time and resources from concrete environmental work.	23	19	28	29
Conflict is a natural part of organizational life.	70	76	64	81
Panel B				
The leadership is right to act on behalf of the organization without always consulting the members.[a]	85	86	84	89
The leadership is good at dealing with conflict within the organization.[a]	80	76	87	81
There is too great a distance between the leadership and the members in the organization.	27	48	39	46
The leadership provides inadequate information on important matters.	23	32	30	50
The leadership acts on its own too often, without considering the members' views.	21	30	23	40
Panel C				
The organization has no business controlling the environmental behavior of its members.[a]	62	67	63	64
There is too little debate and discussion in the organization.	31	43	49	45
Percent "Never"				
Panel D				
Do you get in touch with the organization to take an initiative on important issues?	71	76	61	74
Does the environmental organization contact you to take part in organization work?	57	49	38	59

[a] No statistically significant differences between types of organization at the 0.05 level.
Questions were not posed to the general population
N_{OE} = 681–1982.

There is little organizational difference as to legitimate top-down control in the organization (Table 8.7, Panel C). On the other hand, only members in old noncore organizations seem to want more debate and discussion in the organization. Whether the lack of such debates and discussions in the organization is the reason why so many of them are passive members, we do not know.

Finally, two items tap the level of contact between the organization and the members (Table 8.7, Panel D). Even if the members are not impressively active, the most active members are found in new noncore organizations, where only three in five members never get in touch with the organization. Complementing this, the new noncore organizations most frequently get in touch with their members. Thus, in new noncore organizations, internal communication seem to be most frequent and reciprocal even if the communication is not that comprehensive. Old core members are the most passive group, and the leaderships of these organizations have only to a small extent been in touch with their members on matters on organizational work. The formal democracy in old organizations does not at all imply a high degree of activity and participation in the democratic processes. Still, the formal democracy makes these organizations qualitatively different from the new organizations. In the old organizations, members always had the possibility to influence ideology, leadership, and policies. We observe the ongoing change, but, so far, the knowledge of the long-term consequences for democracy of this important change in organizational form is limited.

All in all, the differences among the four types of organization are small. There does not appear to be distinct differences between members of old and new organizations. Members of the new and nondemocratic type of organization are, in general, more supportive of democracy than what the organizational model itself should indicate. In addition, members of these organizations do not seem to be too critical of their leadership. However, new core members display an interesting pattern. They have the largest proportion of respondents stating that the leadership provides inadequate information and that the leadership acts without considering the views of the members. Simultaneously, they are among those most critical of running the organization in a democratic manner. This brings out an interesting lack of symmetry between the individual level and the organizational level: To a certain extent, the members of new core nondemocratic organizations appear to think as if they were part of a democratically built organization. Members of new core organizations still emphasize the leadership's right to act without consulting its members. The data also convey a tendency of the members of old noncore organizations, who were members who we saw in Chapter 6 were not that active, to be the least critical to their own organization.

Although internal democracy takes time, little evidence in our data suggests that Norwegian environmentalists are willing to abolish democracy in the near future in order to increase the fight for the environment. Democracy has strong support whether or not members are active in the organization. Democracy also has support independently of whether or not the organizational model is democratic.[161] These observations correspond to results obtained for voluntary organizations in general. Within all voluntary organizations, 70% of the members agree that internal democracy is very important to them (Wollebæk et al., 2000). This figure is comparable to the question on whether or not running an organization

[161] In several publications, some of us have analyzed the relationship among organizational form, the role of active versus passive members, and democracy (Selle & Strømsnes, 1998, 2001; Strømsnes, 2001; Wollebæk & Selle, 2002b, 2002c). In these publications, we have argued that the passive members of an organization play a far more important role for civil society at large, including democracy, than what can be read from the contemporary literature on social capital and civil society.

democratically is too costly, in which three in four respondents disagreed. These observations suggest that members of voluntary organizations often find internal democracy important to a greater extent than what modern organizational forms seem to permit. Whether this situation will continue or whether the long-term consequences of the new generation of organizations would mean less emphasis on democracy, we cannot tell. However, our conjecture is that in the long run, more often than not we ought to see symmetry between organizational form and membership attitudes. Therefore, something has to give.

Democracy External to Organizations

From an organizational point of view, external democracy refers to democracy's intrinsic value, efficiency, and fairness. We expect that members in new organizations, which we have classified as being internally nondemocratic, are more inclined to favor environmentalism at the expense of democracy and more inclined to see democracy as inefficient than members in old organizations as well as the general population at large. We expect these differences to be distinct, but not very large. We also expect that the general population attributes greater importance to democracy and defend it as more efficient than do organized environmentalists. This assumption is based on the general environmentalists' view that democracy does not forever hold a privileged position over environmentalism and that democratic means sometimes are insufficient in resolving imminent environmental problems. However, state-friendliness might keep the differences between the general population and organized environmentalists small. Because organized environmentalists also trust political institutions, it will be surprising if they rejected the democratic foundations of these very trustable institutions.

The analysis shows that there is a high level of agreement with the claim that one cannot have an ecologically sustainable society without democracy (see Table 8.8, Panel A). Conversely, there is also a high level of disagreement with the statement that it is more important to solve environmental problems than to secure democratic rights. This indicates a strong inclination toward democratic commitment both across the organizational types and the general population. Indeed, differences between these groups are small or nonexistent. Only members of new core organizations deviate, as they agree more than others that democratic rights should yield to solving environmental problems. When specifically asked about human rights versus environmental protection, only one in six environmentalists advocates human rights. For the general population, the figure is one in four. When democracy in general is pitted against environmental needs, both the general population and organized environmentalists side with democracy. On the other hand, human rights, commonly linked to democracy, has a far more precarious position when considered against environmental protection. Whereas democracy is defended as a general value, one of its tenets, human rights, seems to be an easier hostage. Although

TABLE 8.8. Evaluation of external democracy; percent "agree".

	GP	ONC	OC	NNC	NC
Panel A					
A properly functioning democracy is a prerequisite to the creation of an ecologically sustainable society[a, b]	74	80	75	74	72
It is more important to secure human rights than to protect the environment.[b]	24	18	18	14	14
It is more important to solve environmental problems than to secure democratic rights.	24	22	29	29	36
Panel B					
In the environmental movement, it is sometimes necessary to use illegal actions (civil disobedience) to attract attention.	60	64	68	64	83
In environmental protection it can, in some cases, be absolutely necessary to resort to violence in the most important issues.[a]	9	4	5	10	13
Panel C					
Our democratic system is unfair because the coming generations have no say in environmental matters.	28	34	43	48	51

[a] No statistically significant differences between the general population and organized environmentalists at the 0.05 level.
[b] No statistically significant differences between types of organization at the 0.05 level.
$N_{GP} = 866–966$, $N_{OE} = 1867–1990$.

this is an interesting observation, here we are unable to gauge the depth of this conviction.

When we consider the need for environmental actions against their efficiency of democracy, the analysis shows a distinct difference between civil disobedience and violence (Table 8.8, Panel B). Civil disobedience is endorsed by roughly two in three respondents among the general population and the organized environmentalists. This is in congruence with the situation within the voluntary sector at large (Wollebæk et al., 2000). We are a bit surprised that organized environmentalists did not score higher than the population at large. The exception here is, again, members of new core organizations in which four in five respondents endorse civil disobedience. The necessity of violence only draws support from roughly 1 in 10 respondents. Again, members of new core organizations seem to be most lenient on this issue. On the other hand, members of old organizations are most restrictive. As to efficiency, civil disobedience as an environmental means is broadly legitimized, whereas violence is not.

Finally, on the issue of democratic fairness, which concerns representation of interests for future generations, the general population's view is one of restriction

(Table 8.8, Panel C). Only one in four considers it unfair that coming generations have no say in environmental matters. On the other hand, every member of new organizations agree that such a limitation is unfair. On balance, there is no overwhelming support that it is unfair that the interests of coming generations are absent from our democratic system.[162]

Members of new organizations tend to be more proenvironmentalist and thus less preoccupied with democracy as an absolute value than members of old organizations. This is especially the case for members in new core organizations even if the differences are not that large. Members of old noncore organizations are most positive to external democracy or democracy in society. The pattern of responses indicates that there is a kind of continuum on external democracy (or democracy in society) along which the general population is most democratic and new organizations are most skeptical. Old organizations are found between these two positions. We attribute these small differences on external democracy to the effect of state-friendliness. Despite the small differences among the environmental types, there still seems to be a certain link between organizational structure and members' perception of democracy. New organizations, especially new core organizations, reject internal democracy and also, to some extent, give less priority to external democracy. Here, to some extent, the members of new core organizations follow suit. This corresponds with more general tendencies within the voluntary sector in which new organizations are less often democratically built and new members, especially young ones, less concerned with democracy (Wollebæk & Selle, 2003).

Conclusion

Organized environmentalists' patterns of attitudes toward organizational means, decision-making, democracy, as well as trust in institutions have demonstrated the presence and viability of the state-friendly society. The lack of systematic and major difference in attitudes between the general population and organized environmentalists can only be understood by way of state-friendliness. Without this perspective, it is difficult to make sense of environmentalists and of the environmental movement in Norway.

Conspicuously, members of the environmental movement trust some state institutions more than the general public does. This difference is in addition to Norway's comparable high level of trust in these institutions. This high level of trust might prove to be all in vain if the desire of the environmental movement was to be an alternative movement in opposition to the state. However, the organizations as well as their members do not exist to be in opposition to the state. Their goal is to work

---

[162] This question might be difficult to answer succinctly. What does it really mean that future generations will have a voice? Is it something "metaphysical" or just that politics to a larger extent than we are used to should take a long-term perspective?

for the environmental cause by being almost an integral part of the government. Additionally, we see that the environmentalists are also oriented, either cognitively or ideologically, toward the state rather than being oriented away from it. However, this orientation does not necessarily mean a lack of conflict within or among the organizations or differences in policy orientation between organizations and the state (see also Chapter 3). The ideological orientation is in most, if not all, respects related to the open and inclusive state, with its strong tradition of representative democracy. Having an amicable relationship with the state is a guarantee that the environmental movement in return obtains financial support and influence and receives public legitimacy. These rewards are crucial for visibility in society and for the recruitment of members to the organizations. There is not a large segment of radicals in the general population from which to mobilize. Another generalized reward is the contribution to the plurality and vitality of civil society itself. Contrary to what the Dryzek study (2003) argues, we hold that moderation and state-friendliness are not something that are almost forced down the environmentalists' throats by government bodies. Rather, in a comparative perspective, the overall political culture makes environmentalists and their environmental organizations moderate from the start. To a large extent, there seems to be a symmetry between the members' attitudes and the organizations' positions. Most of the time, environmental organizations are not working against the interests of their members.

The Green Party in Norway failed because the political parties and the government quite early co-opted environmental issues. In the absence of a nuclear question around which environmentalists could rally, the understanding of environmentalism became broad and easily suffused the open political system. Established parties permitted environmental issues to pass through the gates of the party programs only and no access was given to a green party latecomer. This procedure preempted potential political space reserved for a green party. In Norway, the Liberal Party and the Socialist Left Party are functional green parties. However, in the period of environmental stagnation that we are experiencing now, these parties are less green today compared to the situation in the 1980s and early 1990s.

Relatedly, Greenpeace Norway's confrontational and independent approach, such as its antiwhaling campaigns, as well as NOAH—for animal rights' lost cause, collided not only with the Norwegians' state-friendliness but also with the local community tradition of proud and defiant, albeit frugal, self-sufficiency. The organizers behind Greenpeace in Norway failed to gain access to the existing and rewarding political networks that, in return for cooperation, dole out funding from the government. Greenpeace International did not want to sacrifice its stance on whaling. Because they refused to swallow the whale, they did not receive any government pork. The ideology and political strategy of Greenpeace was simply anathema to the general public.

Members of the environmental organizations fail to separate themselves from the general public on issues of democracy's value, efficiency, and fairness. Indeed, on some issues, organized environmentalists are almost indistinguishable from ordinary citizens. Only members of new core organizations have salient attitudes on these issues. So far, however, the differences in attitudes between members of new and old organizations, as well as between democratic or nondemocratic organizations, are not strong enough to be called generational effects. These moderate positions do not vanguards of an alternative movement make. The general impression is one of a rather attitudinal pragmatism around which no deep environmental ideology can be built. Looking at the changing contours of organizational forms in the mid-1980s, we are definitely talking about a generational change in the organizational society clearly expressed within the environmental movement. This has not meant that modern environmentalism has become more of a radical subculture. In the final chapter, we will look a little closer at the possible long-term consequences of the changes already seen in the environmental movement on the two anomalies in this study.

Chapter 9
Withering Uniqueness?

Introduction

We started this study with the claim that the uniqueness of organized environmentalism in Norway stemmed from it combining the anomalies of a state-friendly society with a local community perspective. Throughout this volume, we both analyzed the two anomalies and used them to interpret the results of other analyses. We argued that without understanding these anomalies and how they operate within this particular polity, one is unable to understand organized environmentalism in Norway. Also, one is unable to make sense of the lack of a clear distinction between the general population and organized environmentalists regarding behavior and attitudes. Specifically, in the absence of the anomalies, one is unable to make sense of the fact that members of the Norwegian environmental movement—an alleged alternative movement—outperform the general population regarding institutional trust. These findings should be of interest not only to those who study environmentalism, but also to those who understand the development of social capital primarily as a result of spontaneous bottom-up processes. Our study shows that the role of government and how it relates to the civil society might be crucial in the creation of social capital.[163]

In this comparative case study of the unique environmentalism in Norway we have strongly emphasized the context in which this type of environmentalism is embedded. We argue for the need for such case studies within a comparative context. Such studies permit the researchers to identify and emphasize the deep structure of the political culture that undergirds environmentalism in that country. The political culture plays a crucial role in defining the contents of the environmental organizations in a single polity and the political repertoire that is available for

---

[163] There is a growing literature on the importance of the structure of the welfare state in creating social capital. The literature points out the connection between the universal features of the welfare state and the comparatively high amount of social capital in Scandinavia (e.g., Kumlin & Rothstein, 2005; Rothstein & Stolle, 2003).

their collective action. The more unique the case, the more we need to study it not only to understand the case itself but also to understand the structure and limits of environmentalism in general. We hope that we have been able to show the value of this research strategy.

The way we have organized our study is different from the Rootes study (2003). This study emphasizes the importance of different environmental cultures and of political conjunctures in a single country of which different environmental groups take advantage. We are not arguing against the fact that branches of environmentalism have features in common across countries (e.g., animal rights and antinuclear movements). However, we would not go as far as the Rootes study. We argue that looking at political conjunctures, even if important, does not tell the whole story. Researchers need to understand the broader context in which political conjunctures and environmental culture operate.

Environmentalists and environmental organizations are strongly influenced by the political culture of their home country. Such an approach gives researchers an insight into the structure, the possibilities, and the limits of environmentalism in a given setting. This is especially needed for cases that are qualitatively different from other cases. We have shown that this is the case of Norwegian environmentalism. This case is unique for the environmental movement in that the question of nuclear energy never had the chance to be a political catalyst for a green party and where the animal rights movement never seems to get a pawhold. Both phenomena can be explained by the context and political culture within which environmentalism operates.

We are talking about the importance of anomalies that are a defining part of the political culture itself. If an organization operates too far from the core of the political culture, the organization will never gain any political traction other than in times of great political changes. In a country that is geographically large and where the urban centers are few, small, and far between, there are small chances for political subgroups to build strong bases.[164] If an organization wants to obtain political leverage in such a polity, becoming a political subculture is a cul-de-sac. Rather, the organization should turn toward and not away from the state. That is why we, in this context, often see radical groups advocating their case by getting the attention of the government and convincing it to take "public responsibility" and "take care of" the problem and to pay what it costs.[165]

Furthermore, we have underscored that the Dryzek perspective (2003), even if the authors attempt to answer questions not too different from the ones we have addressed here, contains a somewhat incomplete understanding of the relationship between the state and the civil society in Norway. So far, voluntary organizations are not primarily extensions of the state; nor do they have next to no

---

[164] For a discussion of the importance of size for the probability of establishing strong political subgroups, see Tranvik and Selle (2003).

[165] For instance, this was exactly how the very radical feminist organization The Women's Shelter Movement started its operations; see Morken and Selle (1994).

autonomy in a polity or operate without any grassroot influence. We are not talking about a polity in which there is no vital public space and where the government smothers every local activity. Such a view is out of touch with the historical dynamics of the Norwegian and Scandinavian relationship between the state and the civil society. The reason for not getting it right is a too fixed theory in combination with insufficient knowledge of context. One is unable to get to the core of a phenomenon if one fails to understand the context in which the phenomenon operates. In a comparative perspective, to see Norway as a "thin" democracy, as the Dryzek study (2003) does, is to fail to fully understand how Norwegian and Scandinavian democracies work.[166] This view, together with a lack of emphasis on the important role of local communities, means that the Dryzek study falls short of grasping the core of Norwegian environmentalism.

We have commented on some of the changes that have taken place within the Norwegian polity over the last two decades. These are the new neo-liberal state, the new role of the market sector, the declining importance of the periphery and rural areas more generally in Norwegian politics, the decreased autonomy of local government, and the transformation of the voluntary sector itself.[167] These changes might gradually change the structure of state–civil society relations, exercise pressure on the weight of the local community perspective, and might even strain organizational autonomy. One of the interesting aspects of these changes is that if they continue to develop as they have done over these two decades, they might in not too long a time have a fair chance of in fact proving the Dryzek perspective right. This would mean a development in the direction of a diminished public space in which voluntary organizations increasingly become extensions of the state. In order to further substantiate this conjecture, we need to see where our study of Norwegian environmentalism has taken us so far.

Organized Environmentalists Are Almost Like People in General

In the international literature, young age, higher education, and leftist political attitudes have for a long time been tried and tested correlates of environmental concern. In effect, these correlates should be able to distinguish organized environmentalists from the general population. However, in this study of environmentalism in Norway,

---

[166] Here, we would again like to draw attention to the Johns Hopkins Comparative Nonprofit Sector Project in which more than 40 countries participated. This project has been of great importance in challenging the theoretically based but historical uninformed misunderstanding that a "thin" democracy is the rule in the Scandinavian countries. The project shows that these countries are among the largest and most vibrant civil societies available (Salamon & Sokolowski, 2004).

[167] For an attempt to systematize these changes and their consequences, see Tranvik and Selle (2005) and Selle and Østerud (2006).

we did not detect any significant age differences between the organized environmentalists and the general population. This means that organized environmentalists
in Norway are not at all young compared to the general population. This result is in
part due to the fact that the environmental movement has come of age. However, it
is also due to the fact that environmentalism is less of an alternative movement in
Norway and that it is strongly embedded in the overall political culture. Although not
too many people have become members in these organizations compared to other
types of voluntary organization, the environmental movement has evidently attracted
people of all ages.

Our study confirmed that organized environmentalists have a significantly
longer education than the general population. Also, the parents of organized environmentalists have a higher education level than have parents of the general population. Overall, the picture of members being well educated fits in with what is
observed for the voluntary sector at large. However, the results also show that the
educational level of the members of the environmental movement is somewhat
higher compared to members of the voluntary sector in general. This difference
also seems to increase.[168] The empirical relationship between education and
income is weak but uncontroversial; it follows that organized environmentalists'
higher level of education should also provide them with higher incomes compared with the general population. The distribution of education and income
among environmentalists confirm that Norwegian organized environmentalists
have more cultural and economic capital available than the general population.
These endowments place the environmentalists firmly in the broad category of the
middle class. Again, this is also the situation within the general voluntary sector.
However, nothing in our analyses indicates that new types of organization attract
more resourceful members than old types of organization. Although a new organizational type emerges, it does not consist of members different from the population in general.

Unsurprisingly, members of environmental organizations hold stronger ecological beliefs, they are more postmaterial, and they hold stronger egalitarian values
than the general population. Our study also confirmed the left-leaning of environmentalists through their greater radicalism and preferences for center and leftist
parties. This buttresses the view that the environmentalists prefer to work within
the established party system. This is different from the situation within the voluntary sector at large, in which we find that the parties of the political center are overrepresented. However, our analyses showed that members of core organizations
were politically *more* moderate than members of noncore organizations. This
result leads us to conclude that *environmental coreness is associated with moderate environmentalism*. Furthermore, members of environmental organizations are
more active than the general population in almost any type of political behavior.
This, together with them scoring high on the different trust measures, means that
organized environmentalists are very likely to receive high scores on any social

---

[168] For more discussions on this topic, see Wollebæk and colleagues (2002, 2000b).

capital measure. A large share of members in noncore organizations is active in environmental work and other types of voluntary activity, whereas many members in core organizations are surprisingly passive. These results strongly suggest that members of the single-issue new noncore organizations engage themselves the most in environmental and political behavior. The fact that environmental coreness is associated with moderate environmental attitudes and that members of noncore organizations are most active in environmental work, we argue, are a direct consequence of the strength of the state-friendliness and local community anomalies.

These characteristics lead us to the general conclusion that organized environmentalists are not at all a species different from the general population. Our view is that environmentalists stand with both feet in the dominant and common national culture, they are pragmatic, and they do not hold deep ecology alarmist beliefs. It is unsurprising that organized environmentalists hold stronger environmental attitudes and behave more environmentally friendly than the general population. Any other result would defy wisdom and logic. However, the general results of our study overwhelmingly nail down that organized environmentalists still are an integrated and familiar part of the general public.

Balancing Between State and Local Communities

In Norway, environmental organizations operate in a civil society where there is little room for an organization to present itself as an alternative to the state if the organizations nourish ambitions of achieving political influence.[169] The state provides money through financial support and legitimacy through cooperation. Without the state financial support, the survival of most voluntary organization would be jeopardized unless the organizations scaled their level of organizational activity down to a minimum.[170] However, the organizations' proximity to the government go beyond the organizations' simple need for money. The proximity and cooperation is of a deeper cognitive orientation.

Despite a lack of interest among the younger population, the voluntary sector is trusted among people in general. The sector has a high degree of legitimacy. This legitimacy and trust make it very likely that the environmental organizations in the years to come will continue to cooperate closely with the state. Therefore, there is not much evidence to counter our conclusion that *Norwegians in general, whether they are environmentalists or not, are friendly toward the state.* This state-friendliness is likely to continue even if the economic market has increased

---

[169] With the possible exception of certain religious organizations that are part of the layman's movement, as well as some immigrant communities, there have not been strong alternative movements in Norway since the labor movement was incorporated in the 1930s (see Sivesind et al., 2002; Wollebæk et al., 2000).

[170] However, we would like to emphasize that the Norwegian voluntary sector through membership fees and different types of sale generates more of their own income than in many other Western countries (Sivesind et al., 2002).

its legitimacy during the last two decades and even if some of the environmental organizations have started to cooperate with business organizations. Furthermore, even if the forces of "the new public sector" has transformed the state for some time, results from general population surveys show that, except for trust in parties and politicians, there is no indication of a declining trust in governmental institutions (Listhaug, 2005; Strømsnes, 2003).

Environmental organizations cannot, or see no need to, present themselves as radical alternatives to the state. They might be critical to parts of the policy of changing administrations and the ways in which these are implemented, but we completely failed to find traces of a fundamental skepticism toward the state. If there is any skepticism toward the state at all, it is found among members of new core environmental organizations. Few Norwegians imagine a good society without a strong, open, active, and interventionist state. The state of course plays a core role in environmental politics. However, this does not imply that the environmental movement does not exert significant influence on public policy. For instance, political parties and the state were quick to respond to the environmental movement when it emerged politically in the 1960s: Political parties incorporated environmentalism in their platforms and the Ministry of the Environment was already established in 1972. In the latter case, even the environmental organizations played an important part (Jansen, 1989). However, unlike the Dryzek study (2003), we do not see governmental behavior mainly as a strategy to moderate new and dangerous protest, but as an expression of how politics works in this type of system.

The environmental movement has been important in Norwegian politics. This can only be understood properly by studying how voluntary organizations act in a state-friendly society. The all-encompassing and inclusive Norwegian state has close ties to the civil society and voluntary organizations. The state-friendliness among the citizens legitimizes this relationship. However, the distinction between state and civil society is blurred. There is a large area of cooperation as well as exchange of ideas, information, expertise, and resources between the state and the civil society. The cooperation is close and familiar on a broad level. The organizations are able to influence the state and still maintain their organizational autonomy. Those who fail to see this mutual influence also fail to see the important role of voluntary organizations in this type of society. We are talking about a rather extensive public space in which both governmental bodies and voluntary organizations operate. Indeed, we hold state-friendliness accountable for the fact that *organized environmentalists, more than the general population, trust the national political system and political parties*. However, state-friendliness is more than a question of co-optation. It also simply means that organizations can exert influence on environmental policies.[171]

---

[171] The mechanism that the more your organization works and cooperates with government bodies, the higher the level of institutional trust, is one that we are unable to elaborate upon here. Such an elaboration would include questions on whether trust can be explained by initial trust that is promoted through self-selection or trust that is learned from experience by working with governmental bodies.

The left-leaning of the organized environmentalists does not make it a surprise that *organized environmentalists trust the business sector less than does the general population*. This result is actually what is to be expected from an ostensible alternative movement. However, the observation provides us with a benchmark in the midst of recent development in which both members of new core environmental organizations and the organizations themselves increasingly seem to consider the business community not only as a source of financial contributions but also as an important actor to lobby in order to foster constructive environmental outcomes. If other environmental organizations follow this change of orientation from new core organizations and the business community sees an advantage in such cooperation, a niche of environmental market-friendliness might expand and, in ways difficult to predict, complement the prevailing effect of the state-friendly-society thesis. However, increased legitimacy of market forces might not necessarily weaken state-friendliness. Increased legitimacy would depend on what will happen to the whole relationship among government, market, and civil society in the transformation period we are now witnessing (Tranvik & Selle, 2005).

Our study showed that the organized environmentalists have great confidence in individuals to improve their lifestyles. They also revealed confidence in local solutions to environmental problems. These observations illustrate the weak international orientation of environmental organizations in general, which is a point that is also supported by the Rootes (2003) study that the nation state, not the international community, still is the main environmental point of gravity. The Rootes study, however, is concerned with actual political protest and says little about the cognitive orientation of the environmentalists in the different countries. Our observations of the Norwegian environmentalists add to the Rootes study that *within the nation-state, environmentalists are oriented toward local communities*. In addition to this view being held by the organized environmentalists, the organizations also hold this view despite the fact that they have become increasingly centralized organizations and mainly operate at the national level (Strømsnes, 2001). Despite the simple fact that all environmental problems cannot be solved locally, the organizations' local anchoring is evidence of a strong ideological and cognitive component of the national culture.

A consequence of the local community perspective is the failure to incorporate animal rights in the definition of environmentalism.[172] Our survey shows that animal rights, in the meaning of conflict with the local community per-

---

[172] We are not ignoring the fact that animal welfare receives strong support in Norway. As argued in Chapter 7, it is important to distinguish between, on the one side, a general prevention of pain and suffering as well as a decent treatment of domesticated animals and, on the other side, animals that humans have a right to hunt as well as predators that challenge the local community doctrine of protection of man in nature. This, of course, does not rule out the combination that one can strongly support animal welfare without necessarily defining oneself as an environmentalist.

spective, is endorsed neither by the general population nor by organized environmentalists. In addition, leaders of the environmental organizations, in our interviews with them, underscored this lack of endorsement of animal rights (Strømsnes, 2001). Unsurprisingly and consistent with the goals of their organizations, members of NOAH—for animal rights and, to a lesser extent, members of Greenpeace, evaluate animal rights as more important than do members of any other organization in our survey. It follows that members of NOAH and Greenpeace are less oriented toward the local communities than are members of the other organizations. In general, if there is a conflict between local communities and animal rights, both the general population and organized environmentalists, with the exception of members of the two said organizations, strongly side with the traditional interests of the local communities. *A conspicuous and important effect of the two anomalies is that Greenpeace, one of the most important international environmental organizations, finds it so hard to establish a bridgehead in Norway.*

In our analyses, among organized environmentalists we found that generational distance to a farm is an important and robust variable that explains animal rights attitudes and, by implication, the strength of the local community perspective. *Thus, the greater the distance to a farm, the stronger the endorsement of animal rights.* Interestingly enough, we did not find this relationship among the population at large. Although distance to a farm shows the effect of the local community perspective, it is, given the ongoing urbanization, also an effect that will wane as years go by. Should the level of urbanization in Norway gradually reach continental levels, Norwegian environmentalists will probably become less anomalous and more similar to environmentalists in other countries. At the same time, they will become increasingly different from the Norwegian population at large. Even so, in the longer run it would probably also mean that the attitudes of the general population will gradually change. Therefore, urbanization is an antidote to what we argue is the effect of local community on environmentalism. However, what we have addressed and observed here, and in more ways than can be measured by surveys, we hold that, at least for now, *the local community perspective has had a profound influence on organized environmentalists, their organizations, as well as the population at large.* The local community perspective is, so far, still deeply embedded in the overall political culture.

We stand by the effect of local community on Norwegian environmentalism. The thin end of the wedge against our local community argument consists of two aspects. One is the degree to which the local community perspective is properly measured through the animal rights index. For further research, we, of course, prefer to see that this index will be confirmed by others. However, we would also welcome other constructs that tap the effect of local community. The other aspect is that a waning of local community, in whatever way it is measured, can bring about unpredictable effects. Although we do not observe this waning yet, once this strong cognitive and cultural dimension diminishes we will be unable to foresee the full range of consequences and interactions that will occur.

We complete the empirical analysis of our survey by placing the 12 organizations and the general population within the 2 anomalies. We use the animal rights index as a measure of *the local community perspective*. Here, higher scores are associated with a greater commitment to local community. Also, we use the index of trust in institutions as a measure of *the state-friendly society*. Higher scores are associated with a greater degree of trust.[173] Figure 9.1 shows the mean scores of the 13 groups on the 2 dimensions.

The pattern shows that both NOAH and Greenpeace deviate significantly from all environmental organizations and the general population in Norway (marked with the solid bullet). The two organizations are found to combine a moderate state skepticism with a strong rejection of the local community perspective. It is interesting to note that the general population and the 10 groups (other than NOAH and Greenpeace) primarily show variance along the state-friendliness dimension but not the local community dimension. On the local community dimension, the 10 organizations show a high degree of consensus as well as agreement with the population at large. NOAH and Greenpeace deviate from this pattern in that they also show variance along the local community perspective. Overall, Figure 9.1 underscores the deviance of NOAH and Greenpeace in Norwegian

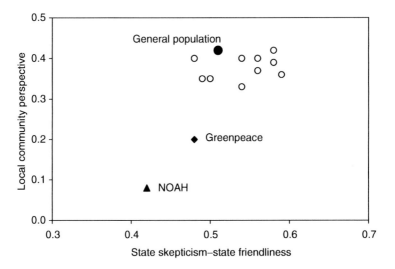

FIGURE 9.1. The local community perspective and state friendliness. $N_{GP} = 880–939$, $N_{OE} = 1823–1842$.

---

[173] For a discussion on the local community perspective and the coding of the index, see Chapter 7. The state-friendly society index is based on the three items of trust: environmental authorities, political system, and political parties, see Table 8.6.

environmentalism. It shows quite clearly the political distance that they have established between themselves and the environmental mainstream, as well as the population at large.

Second, we also seek to identify another pattern between the two dimensions by taking the mean scores on the local community perspective for every category of state-friendliness (see Figure 9.2). This pattern will show the degree of local community perspective given the level of state-friendliness. The graph shows that for the general population, there is hardly any relationship between state-friendliness and the local community perspective. The correlation is insignificant (r_{xy} is −0.06). For organized environmentalists, the conclusion is more interesting. When state-friendliness increases, the local community perspective also gains ground. Inversely, when state-friendliness decreases, the local community perspective decreases. This relationship is moderate, but significant. The correlation (r_{xy}) is 0.20. Basically, this finding suggests that organized environmentalists, to a greater extent than the general population, are influenced by both anomalies. These results strengthen our claim that *it is necessary to understand both the local community perspective and the state-friendly society in order to understand the anomalous case of organized environmentalism in Norway*. However, how long will this conclusion be valid?

Withering Uniqueness?

The two final decades of the 20th century brought about distinct changes, all of which brought to bear, and will continue to bear, on environmental organizations and voluntary organizations in general. These changes affect the structure and

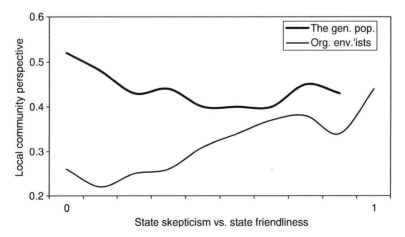

FIGURE 9.2. The local community perspective and state friendliness.
$N_{GP} = 823$, $N_{OE} = 1650$. *Note*: Curves have been smoothed.

internal life of the organizations, the relationship among organizations, as well as the relationship among state, market, and the voluntary sector.

One important change is the neo-liberal shift in Western societies in which the market and market mechanisms received more prominent positions and gained unprecedented legitimacy in modern times. Large and important sectors in society, for instance credit and housing, became depoliticized when governance and audit was transferred from the state to the logic of the market. The general increase in market orientation and the market's legitimacy work their way into the public sector and, in turn, affect the ways in which the public sector is organized and how it relates to other actors. The market liberalization has reached and suffused most areas of society, including environmental organizations. This influence is not merely a matter of economic management in the voluntary sector. It is also a sign of a closer relationship between the voluntary sector and the market sector in that the degree of cooperation between the two has increased. However, what the market can offer is not necessarily what traditional membership-based organizations need. If one implements too much market logic and merely treats members as customers, regular membership, as we know it, might vanish. Two consequences are that the organizations might become increasingly centralized and professionalized. The development within the environmental movement since the mid-1980s can be seen as a prototypical example of this more general transformation that now takes place in the voluntary sector. The roles of the environmental organizations as democratic agents and intermediaries are therefore put under considerable pressure.

Furthermore, the profound development of information and communication technologies influences how individuals organize themselves in the society. New technologies also affect the structures and behaviors of traditional organizations. An organization is a kind of communication system. Once new technology takes hold of an organization, it might alter the ways in which the organization operates.[174] Similarly, we observe the waning role of traditional membership in organizations founded in the period after 1985. These new organizations are more professionalized and more centralized. They are less democratically built and they hardly base their activities on members at all, even though we have seen that many supporters of these organizations are very active both within their organization and in general. This organizational change is so important that it has been a constituent for our organizational typology.

Almost all 12 organizations in our study have set up homepages on the Internet available for both their members and the general public. Many organizations also ask specifically for feedback and input from those who visit their sites. These sites are continuously improved and updated.[175] The proliferation and importance in the information and communication technologies challenge these voluntary

---

[174] On the ways in which organization and technology is interwoven, see Douglas (1982), Thompson (2000), and Tranvik and colleagues (2000, 2003).

[175] There is a website for all organizations in our survey, with the exception of Women–Environment–Development (see Appendix B).

organizations in several ways. This is so because the new technology alters the definition and understanding of speed, distance, and efficiency. One can foresee that organizations will increasingly use the Internet-, electronic mail-, and cell phone-related technologies as their primary media of communication. The organization can update members on a continuous basis and they can receive input and feedback from members about current issues.

The comprehensive information systems within the organizations traditionally include newsletters, newspapers, and magazines. The information systems are primarily constructed so that they provide not only information regarding the work of the organization but also important information regarding the area in which the organization is active. All of the core environmental organizations in Norway have extensive information systems that demonstrate an impressive standard and breadth. The new types of organization were among the first organizations to take full advantage of information and communication technologies (ICT). ICT opens up more interactive communication between members and leaders of the organizations, probably making the need for traditional membership meetings less necessary.[176] In addition, the members are able to communicate efficiently among themselves either on a one-to-one or a one-to-many basis. The organization's agenda can, therefore, on short notice, be challenged by members who, most of the time, previously found themselves only on the receiving end of organizational activities. Such communications can complement, if not replace, traditional meetings. The signs of such changing venues in the voluntary sector are already well under way (Wollebæk & Selle, 2002b).

These new ways of organizing also raise the issue of interest representation within organizations more broadly because one already observes such changes taking place in many other parts of the voluntary sector. Over the last few years, the number of membership meetings in the traditionally built organizations has actually decreased while the number of board meetings has increased. This is a clear indication of increased centralization and professionalization. Information and communication technologies will continue to challenge voluntary organizations. They will need to assess closely in what ways such technologies can be used both to maintain and develop an efficient organization, keep its distinctiveness, and also in what ways they can be used to promote the goals of the organization. For instance, to what extent does the new technology fit the new organizational forms better than the more traditional ones? Is the technology flexible enough to strengthen the communication and coherence within more traditional and hierarchically based organizations? From what can be seen not only from studies of the environmental movement but also from the

---

[176] Many researchers understand the general decline in face-to-face contact as one of the main reasons for the decline in social capital in Western countries (e.g., Putnam, 2000; Stolle & Hooge, 2004). Other researchers argue against putting too much emphasis on the importance of face-to-face contact (e.g., Selle & Strømsnes, 2001; Wollebæk & Selle, 2002c).

voluntary sector in general, the more hierarchically built organizations use the new technology to the same extent as the new types of organization. However, the long-term consequences of this profound change in communication are still too early to call.[177]

The neo-liberal ideology has boosted the legitimacy of the market, trade, and business. The ideology has also transformed the way in which the government operates. At the same time, we have observed an increased interest in the civil society and its organizations. Together, this calls for a new understanding of the state–market–civil society triptych.[178] Increased market orientation and the rise of the new technology, among other factors, have led to a shift in the understanding of organizational efficiency and success in many organizations. The old under-standing included high membership numbers and presumed active members (even though many members were rather passive). In the largest environmental organizations, the proportion of active members has always been low. The new understanding often includes passive supporters who contribute money through the membership fee (and other types of contribution) but fail to show up for activ-ities on behalf of the organization. Nevertheless, even without a high turnout of its membership when compared with other voluntary organizations, it is still undisputable that the environmental movement has had quite an impact on public policy-making.

Furthermore, our perception of supporters, including passive ones, depends on whether one emphasizes the internal or the external democratic functions of an organization. If one emphasizes the external democratic role of the organization in democracy, the question of membership activity and internal democratic rights will be less important. In this case, it might be sufficient that others simply rep-resent the views of the members or supporters. Even if this is the situation, the organization itself might still play an important democratic role in society.

However, it might be the case that members of the new organizations are just as committed to the environmental cause and no less attached to their organiza-tions than are members in traditional organizations. This might go for both active and passive members. We have not observed strong and systematic differences between the members of new and old organizations as to the level of activity and commitment. Members of new organizations are at least as active as those mem-bers of old organizations. This should not have been the case if members of old organizations were more committed to the cause. We conclude that the new form of organizing environmentalism involves more than members just donating a meager sum of money to a good cause in order to ease a guilty conscience. New

---

[177] For further disussions on these topics, see Wollebæk and Selle (2002a, 2002b).

[178] The corporatist perspective (i.e., the close and formalized relationship among govern-ment, business organizations, and labor, which often has been employed to explain the Scandinavian countries) has been weakened (Espeli, 1998; Nordby, 1994; Rommetvedt, 2000). The international corporatism literature is comprehensive (e.g., Schmitter,1979; Williamson, 1989). Norway is ranked second on an index of corporatism, only bypassed by Austria (Lijphart & Crepaz, 1991). See also Siaroff (1999).

types of organizations can have a crucial impact on the structure of civil society (Jordan & Maloney, 1997; Selle & Strømsnes, 2001).

In a time when interest for traditional politics is going down and a smaller part of the voluntary sector remains political in character, we observe a decline of voluntary organizations as important political intermediaries. However, in this picture of overall political decline, the environmental organizations, together with the feminist movement, are perhaps the most important of the new politically oriented organizations that have appeared during the last 25 years. We add to this observation our finding in the study that organized environmentalists are politically more active than the general population. Even if the voluntary sector changes, politics still remains salient in the environmental field.

Regardless of organization or affiliation, being an environmentalist influences the way one organizes one's life. Combined with the middle-class status of organized environmentalists in general, this does not at all indicate that members of new organizations are marginalized compared to members of more traditional organizations. True, members of new organizations have a lower level of education and less income than members of old organizations. However, members of new noncore organizations are found to be the most active and ideologically most conscious members of the environmental movement. New organizations also have a small but active share of members and supporters who devote much time and money to environmental work. These results give us few reasons to believe that members and supporters of new organizations simply are passive and powerless supporters (Strømsnes, 2001; Wollebæk & Selle, 2002c). In many ways, these new organizations seem to work as more than "protest-businesses" or "checkbook" organizations (Jordan & Maloney, 1997), even if that appears to be their organizational form. The new organizations seem to be more important to the members and supporters than what can be read from the new organizational model itself. The individual motivation seems to be stronger than what the new organizational model requires. So far, it seems to be a lack of symmetry between organizational form and individual motivation. The discrepancy in the relationship between an organization and its members evidently calls for another study.

Additionally, even passive supporters contribute to organizational finances and, therefore, to the survival of their organizations. This is important for the visibility of the movement in the society at large. In so doing, the fee-paying members are important in sustaining the plurality of civic society. In other words, passive supporters are important for the structure of civil society itself and, by implication, for the citizens who benefit from a plural society but are not directly connected to civil society organizations. Therefore, there is no reason to subscribe to the view that passive supporters contribute next to nothing to civil society and democracy. Therefore, passive members or supporters are not necessarily guilty of contributing to the decay of democracy.[179]

---

[179] See Putnam (2000) and Wollebæk and Selle (2002b, 2002c) for different positions on this debate.

All in all, the environmental movement—as well as the voluntary sector in general and the population at large—has been influenced by the local community and state-friendliness anomalies. This influence has contributed to give the Norwegian environmental movement a sense of unity and prevented their members from standing out from the population at large. The members have not constituted a separate political culture. If new organizations nibble away at the local community perspective and state-friendliness, Norwegian environmentalism might become less similar to the common and consensual Norwegian way of life, more ideologically diverse, and more similar to their counterparts in the other countries that we have held up for comparisons. However, even if that might mean a more radical environmental movement with a broader repertoire of collective action, that would not necessarily mean stronger influence on environmental thinking and policies.

The changes that we now see in the public sector, brought about by the New Public Management syndrome, have a profound impact on how the state interacts with other sectors and institutions. The changes affect the relationship between government bodies and voluntary organizations. Increasingly, government bodies implement policies through "contracts" with other actors in a system of strong control, emphasizing efficiency and cost-effectiveness. When implementing public policies, the government system now imposes increased professionalization and centralization when cooperation with organizations takes place. Within the system of stronger control and audit of how government money is used, organizations really have to know the system and their way about it in order to do things right. If this development continues in the years to come, it might make the co-optation argument of the Dryzek study increasingly relevant. The development might gradually transform many voluntary organizations into becoming "arms of the state" and turning them into organizations with no or little autonomy. Such organizations will turn into implementers of public policies rather than crucial partners in the decision-making process itself. If this scenario turns out to be the case, and it is not too unlikely that it will, it is tempting to draw the conclusion that *the pressure of the "neo-liberal state" on organizational autonomy today seems to be stronger and more consequential than what took place under heydays of social democracy and the universal welfare state.*

As part of this transformation, we also see that the center–periphery relation is changing, in that what is urban becomes more important both cognitively and as a political reality. This is followed by a decrease in the autonomy of local municipalities. This process would weaken one of the main institutions that have been so important in keeping the local community perspective going. Furthermore, this happens at the same time as we see a transformation of the voluntary sector, in which a more dual and less integrative voluntary sector appears. We see a decline in the historically important role of voluntary organizations in which they integrated citizens across the local, regional, and central geographical levels. The withering of this vertical integration is gradually replaced with more centralized and professionalized organizations. For the organizations, this development means less emphasis on membership and a marginalization of internal democracy. Members might not share this shift of emphasis.

Even if we can observe tendencies that point in the direction of a weakening of the local community perspective and profound changes in the relationship between state and civil society, we have not observed such a regime change yet. However, the gradual transformations that now take place make it more interesting than ever to follow Norwegian environmentalism and the more general developments in civil society. Insight into these processes should be of interest not only to those interested in Norwegian and Scandinavian politics but also to those interested in the overall conditions of and change in environmentalism and civil society in Western societies.

References

Aardal, B. (1990). Green Politics: A Norwegian Experience. *Scandinavian Political Studies, 13*(2), 147–164.

Aardal, B. (1993). *Energi og Miljø. Nye Stridsspørsmål i Møte med Gamle Strukturer.* Oslo: Institutt for samfunnsforskning.

Aardal, B., & Valen, H. (1995). *Konflikt og Opinion.* Oslo: NKS-forlaget.

Adams, C. J. (1990). *The Sexual Politics of Meat. A Feminist–Vegetarian Critical Theory.* Cambridge: Polity Press.

Ajzen, I., & Fishbein, M. (1977). Attitude–Behavior Relations: A Theoretical Analysis and Review of Empirical Research. *Psychological Bulletin, 84*, 888–918.

Båtstrand, S. (2005). Store Svensker, Små Nordmenn. Grønne Partiers Ulike Grader av Valgsuksess. Paper. Department of Comparative Politics: University of Bergen.

Berenguer, J., Corraliza, J. A., & Martín, R. (2005). Rural–Urban Differences in Environmental Concern, Attitudes, and Actions. *European Journal of Psychological Assessment, 21*(2), 128–138.

Bergens Tidende, (1999, January 31). Fremtidskloden. *Bergens Tidendes månedsmagasin.*

Berger, P., & Neuhaus, J. (1977). *To Empower People: The Role of Mediating Structures in Public Policy.* Washington, DC: American Enterprise Institute.

Berglund, S., & Lindstrøm, U. (1978). *The Scandinavian Party System(s).* Lund: Studentlitteratur.

Berntsen, B. (1994). *Grønne Linjer. Natur-og Miljøvernets Historie i Norge.* Oslo: Grøndahl Dreyer/Norges Naturvernforbund.

Berven, N., & Selle, P. (Eds.) (2001). *Svekket Kvinnemakt? De Frivillige Organisasjonene og Velferdsstaten.* Oslo: Gyldendal Akademisk.

Bjerke, S. (2001). The Sophie Prize 2001. http://odin.dep.no/odinarkiv/norsk/dep/md/2001/annet/022001-210097/index-dok000-b-n-a.html [July 1, 2005].

Bjerkli, B., & Selle, P. (2003). *Samer, Makt og Demokrati. Sametinget og den Nye Samiske Offentligheten.* Oslo: Gyldendal Akademisk.

Bjørklund, T., & Hellevik, O. (1988). De Grønne Stridsspørsmål i Norsk Politikk. *Politica, 20*(4), 414–431.

Bomann-Larsen, T. (1995). *Roald Amundsen: En biografi.* Oslo: Cappelen.

Bortne, Ø., Grendstad, G., Selle, P., & Strømsnes, K. (2001). *Norsk Miljøvernorganisering Mellom Stat og Lokalsamfunn.* Oslo: Samlaget.

Bortne, Ø., Selle, P., & Strømsnes, K. (2002). *Miljøvern Uten Grenser?* Oslo: Gyldendal Akademisk.

Botvar, P. K. (1996). "En Ny Himmel og en Ny Jord." Religion, Politikk og Holdninger til Miljøvern. *Norsk Statsvitenskapelig Tidsskrift, 12*(2), 141–161.

Bramwell, A. (1989). *Ecology in the 20th Century*. New Haven, CT: Yale University Press.

Bramwell, A. (1994). *The Fading of the Greens. The Decline of Environmental Politics in the West*. London: Yale University Press.

Bratland, E. (1995). Sivilt Samfunn og Nye Sosiale Bevegelser. Bergen: LOS-senteret. Report 9533.

Brox, O. (1966). *Hva Skjer i Nord-Norge?* Oslo: Pax.

Brundtland G. H. et al. (1987). *Our Common Future. World Commission on Environment and Development*. Oxford: Oxford University Press.

Brunsson, N., & Olsen, J. P. (Eds.) (1990). *Makten at Reformera*. Stockholm: Carlssons.

Castello-Cortes, I. (Ed.) (1994). *World Reference Atlas*. Oslo: Teknologisk Forlag.

Catton, W. R. J., & Dunlap, R. E. (1980). A New Ecological Paradigm for Post-Exuberant Sociology. *American Behavioral Scientist, 24*(1), 15–47.

Clarke, H. D., Kornberg, A., McIntyre, C., Bauer-Kaase, P., & Kaase, M. (1999). The Effect of Economic Priorities on the Measurement of Value Change: New Experimental Evidence. *American Political Science Review, 93*(3), 637–647.

Cohen, J. L., & Arato, A. (1992). *Civil Society and Political Theory*. Cambridge, MA: MIT Press.

Coleman, D. A. (1994). *Ecopolitics. Building a Green Society*. New Brunswick, NJ: Rutgers University Press.

Dalton, R. (1988). *Citizen Politics in Western Democracies*. Chatham, NJ: Chatham House.

Dalton, R. J. (1994). *The Green Rainbow. Environmental Groups in Western Europe*. New Haven, CT: Yale University Press.

Dammann, E. (1979). *Revolusjon i Velstandssamfunnet*. Oslo: Gyldendal.

Daunton, M., & Hilton, M. (Eds.) (2001). *The Politics of Consumption. Material Culture and Citizenship in Europe and America*. Oxford: Berg Publishers.

Davis, D. W., & Davenport, C. (1999). Assessing the Validity of the Postmaterialism Index. *American Political Science Review, 93*(3), 649–664.

Dekker, P., & van den Broek, A. (1998). Civil Society in Comparative Perspective: Involvement in Voluntary Associations in North America and Western Europe. *Voluntas, 9*(1), 11–38.

Dietz, T., Kalof, L., & Stern, P. C. (2002). Gender, Values, and Environmentalism. *Social Science Quarterly, 83*(1), 353–364.

Dobson, A. (1990). *Green Political Thought*. London: Routledge.

Dobson, A. (1993). Ecologism. In R. Eatwell & A. Wright (Eds.), *Contemporary Political Ideologies* (pp. 216–238). London: Pinter Publishers.

Dobson, A., & Lucardie, P. (Eds.) (1993). *The Politics of Nature: Explorations in Green Political Theory*. London: Routledge.

Doherty, B., & de Gaus M. (Eds.) (1996). *Democracy and Green Political Thought: Sustainability, Rights and Citizenship*. London: Routledge.

Donahue, M. J. (1993). Prevalence and Correlates of New Age Beliefs in Six Protestant Denominations. *Journal for the Scientific Study of Religion, 32*(2), 177–184.

Douglas, M. (1966). *Purity and Danger. An Analysis of Concepts of Pollution and Taboo*. London: Routledge/Kegan Paul.

Douglas, M. (1972). Environments at Risk. In J. Benthall (Ed.), *Ecology: The Shaping Enquiry* (pp. 129–145). London: Longman.

Douglas, M. (1982). Cultural Bias. In M. Douglas (Ed.), *In the Active Voice*. London: Routledge/Kegan Paul.

Douglas, M., & Wildavsky, A. (1982). *Risk and Culture. An Essay on the Selection of Technical and Environmental Dangers*. Berkeley: University of California Press.

Dryzek, J. S. (1993). "Environmentalism and Political Theory" by Robyn Eckersley. *American Political Science Review, 87*(3), 765.

Dryzek, J. S. (1996). Political Inclusion and the Dynamics of Democratization. *American Political Science Review, 90*(1), 475–487.

Dryzek, J. S., Downes, D., Hunold, C., Schlosberg, D., & Hernes, H.-K. (2003). *Green States and Social Movements. Environmentalism in the United States, United Kingdom, Germany and Norway.* Oxford: Oxford University Press.

Dunlap, R. E. (1975). The Impact of Political Orientation on Environmental Attitudes and Actions. *Environment and Behavior, 7*(4), 428–454.

Dunlap, R. E. (1995). Public Opinion and Environmental Policy. In J. P. Lester (Ed.), *Environmental Policy: Theories and Evidence* (pp. 63–114). Durham, NC: Duke University Press.

Dunlap, R. E., & Van Liere, K. D. (1978). The "New Environmental Paradigm": A Proposed Measuring Instrument and Preliminary Results. *Journal of Environmental Education, 9*(4), 10–19.

Dunlap, R. E., Van Liere, K. D., Mertig, A. G., & Jones, R. E. (2000). Measuring Endorsement of the New Ecological Paradigm. A revised NEP scale. *Journal of Social Issues, 56*(3), 425–442.

Eckersley, R. (1989). Green Politics and the New Class: Selfishness or Virtue? *Political Studies, 37*, 205–223.

Eckersley, R. (1992). *Environmentalism and Political Theory. Toward an Ecocentric Approach.* London: UCL Press.

Eckstein, H. (1997). Social Science as Cultural Science, Rational Choice as Metaphysics. In R. J. Ellis & M. Thompson (Eds.), *Culture Matters: Essays in Honor of Aaron Wildavsky* (pp. 21–44). Boulder, CO: Westview Press.

Ellis, R. J., & Thompson, F. (1997a). Culture and the Environment in the Pacific Northwest. *American Political Science Review, 91*(4), 885–897.

Ellis, R. J., & Thompson, F. (1997b). Seeing Green: Cultural Biases and Environmental Preferences. In R. J. Ellis & M. Thompson (Eds.), *Culture Matters: Essays in Honor of Aaron Wildavsky* (pp. 169–188). Boulder, CO: Westview Press.

Endal, D. (1996). Miljøheimevernet: Fra Tilskuere til Deltakere. In K. Strømsnes & P. Selle (Eds.), *Miljøvernpolitikk og Miljøvernorganisering mot år 2000* (pp. 233–257). Oslo: Tano Aschehoug.

Espeli, H. (1998). *Skattereform og Lobbyvirksomhet. En Case-Studie av Stortingets Behandling av skattereformens Delingsmodell (1990–95) og Forsøkene På å Påvirke Utfallet av Denne.* Oslo: Institutt for samfunnsforskning.

Esty, D., & Cornelius, P. K. (Eds.) (2002). *Environmental Performance Measurement: The Global Report 2001–2002.* Oxford: Oxford University Press.

Esty, D. C., Levy, M., Srebotnjak, T., & Sherbinin, A. D. (2005). *2005 Environmental Sustainability Index: Benchmarking National Environmental Stewardship.* New Haven, CT: Yale Center for Environmental Law & Policy.

Fardon, R. (1999). *Mary Douglas. An Intellectual Biography.* London: Routledge.

Ferry, L. (1996). *Ny Økologisk Orden. Treet, Dyret og Mennesket.* Oslo: Tiden Norsk Forlag.

Flanagan, S. C. (1987). Value Change in Industrial Societies. *American Political Science Review, 81*(4), 1303–1319.

Flora, P., Kuhnle, S., Urwin, D. (Eds.) (1999). *State Formation, Nation-Building, and Mass Politics in Europe. The Theory of Stein Rokkan.* Oxford: Oxford University Press.

Franklin, M. N., & Rüdig, W. (1995). On the Durability of Green Politics. Evidence from the 1989 European Election Study. *Comparative Political Studies, 28*(3), 409–439.

Freudenburg, W. R. (1991). Rural–Urban Differences in Environmental Concern. A Closer Look. *Sociological Inquiry, 61*(2), 167–198.

Freudenburg, W. R., & Gramling, R. (1989). The Emergence of Environmetal Sociology: Contributions of Riley E. Dunlap and William Catton, Jr. *Social Inquiry, 59*(4), 439–452.

Fuchs, D., & Klingemann, H.-D. (1989). The Left–Right Schema. In M. K. Jennings, J. W. van Deth, S. H. Barnes, D. Fuchs, F. J. Heunks, R. Inglehart, M. Kaase, H.-D. Klingemann, & J. J. A. Thomassen (Eds.), *Continuities in Political Action. A Longitudinal Study of Political Orientations in Three Western Democracie* (Vol. 5, pp. 203–234). Berlin: Walter de Gruyter.

Gilljam, M., & Oscarsson, H. (1996). Mapping the Nordic Party Space. *Scandinavian Political Studies, 19*(1), 25–43.

Gleditsch, N. P., & Sverdrup, B. O. (2002). Democracy and the Environment. In E. Page & M. R. Redclift (Eds.), *Human Security and the Environment. International Comparisons*. Cheltenham, UK: Edgar Elgar.

Greenpeace. (1995). *Annual Report 1995*. Amsterdam: Greenpeace International.

Greenpeace Norden. (2000). http://www.greenpeace.se/templates/template_37.asp? http://www.greenpeace.no/1aboutus/11main.htm?lang=22 [June 27, 2000].

Grendstad, G. (1999). The New Ecological Paradigm Scale. Examination and Scale Analysis. *Environmental Politics, 8*(4), 194–205.

Grendstad, G. (2003a). Comparing Political Orientations. Grid–Group Theory Versus the Left–Right Dimension in the Five Nordic Countries. *European Journal of Political Research, 42*(1), 1–21.

Grendstad, G. (2003b). Reconsidering Nordic Party Space. *Scandinavian Political Studies, 26(3)*, 193–217.

Grendstad, G., & Ness, N. T. (2006). Now You See It. Now You Don't. The Norwegian Green Party. In W. Rüdig (Ed.), *Green Party Members in Western Europe*. Cambridge, MA: MIT Press.

Grendstad, G., & Selle, P. (1995). Cultural Theory and the New Institutionalism. *Journal of Theoretical Politics, 7*(1), 5–27.

Grendstad, G., & Selle, P. (1999). The Formation and Transformation of Preferences. Cultural Theory and Postmaterialism Compared. In M. Thompson, G. Grendstad, & P. Selle (Eds.), *Cultural Theory as Political Science*. London: Routledge.

Grendstad, G., & Selle, P. (2000). Cultural Myths of Human and Physical Nature: Integrated or Separated? *Risk Analysis, 20*(1), 29–41.

Grendstad, G., & Strømsnes, K. (1996). A Green Polity Without a Green Party: Where Have All the Voters Gone? *Paper*. Department of Comparative Politics: University of Bergen.

Grendstad, G., & Wollebæk, D. (1998). Greener Still? An Empirical Examination of Eckersley's Ecocentric Approach. *Environment and Behavior, 30*(5), 653–675.

Grubb, M. (2001). Relying on Manna from Heaven? *Science*, *294*: 1285–1287.

Guldberg, T.-I., & Schandy, T. (1996). WWF Verdens Naturfond: Fra Panda til Bærekraftig Utvikling. In K. Strømsnes & P. Selle (Eds.), *Miljøvernpolitikk og Miljøvernorganisering mot år 2000* (pp. 146–156). Oslo: Tano Aschehoug.

Gulichsen, K. (1996). Kvinner og Miljø: En Organisasjon uten Framtid? In K. Strømsnes & P. Selle (Eds.), *Miljøvernpolitikk og Miljøvernorganisering mot år 2000* (pp. 228–232). Oslo: Tano Aschehoug.

Gundelach, P. (1993). The New Social Movements in the Nordic Countries. In T. Boje & S. E. O. Hort (Eds.), *Scandinavia in a New Europe* (pp. 337–364). Oslo: Scandinavian University Press.

Gundelach, P., & Torpe, L. (1997). Social Reflexivity, Democracy and Citizen Involvement. In J. V. Deth (Ed.), *Private Groups and Public Life. Social Participation, Voluntary Associations and Political Involvement in Representative Democracies.* London: Routledge.

Gundersen, F. (1991). Utviklingstrekk ved Miljøbevegelsen i Norge. *Sosiologi i Dag, 2,* 12–35.

Gundersen, F. (1996). Framveksten av den Norske Miljøbevegelsen. In K. Strømsnes & P. Selle (Eds.), *Miljøvernpolitikk og Miljøvernorganisering mot år 2000* (pp. 37–81). Oslo: Tano Aschehoug.

Guth, J. L., Green, J. C., Kellstedt, L. A., & Smidt, C. E. (1995). Faith and the Environment: Religious Beliefs and Attitudes on Environmental Policy. *American Journal of Political Science, 39*(2), 364–382.

Habermas, J. (1987). *Lifeworld and System: A Critique of Functionalist Reason* (Vol. 2). Boston: Beacon Press.

Hallin, P. O. (1995). Environmental Concern and Environmental Behavior in Foley, a Small Town in Minnesota. *Environment and Behavior, 27*(4), 558–578.

Haltbrekken, L. (1996). Natur og Ungdom: Dypøkologisk og Antroposentrisk? In K. Strømsnes & P. Selle (Eds.), *Miljøvernpolitikk og Miljøvernorganisering mot år 2000* (pp. 157–162). Oslo: Tano Aschehoug.

Held, D. (1996). *Models of Democracy* (2nd ed.). Cambridge: Polity.

Holland, R. W., Verplanken, B., & Van Knippenberg, A. (2002). On the Nature of Attitude–Behavior Relations: The Strong Guide, the Weak Follow. *European Journal of Social Psychology, 32*(6), 869–876.

Hunter, L. M., Hatch, A., & Johnson, A. (2004). Cross-National Gender Variation in Environmental Behaviors. *Social Science Quarterly, 85*(3), 677–694.

Huntford, R. (1993). *Scott and Amundsen.* London: Orion Publishing.

Huntford, R. (2002). *Nansen: The Explorer as Hero.* London: Abacus.

Inglehart, R. (1971). The Silent Revolution in Europe: Intergenerational Change in Post-industrial Societies. *American Political Science Review, 65*(4), 991–1017.

Inglehart, R. (1977). *The Silent Revolution: Changing Values and Political Styles among Western Publics.* Princeton, NJ: Princeton University Press.

Inglehart, R. (1979). Value Priorities and Socioeconomic Change. In S. Barnes & M. Kaase (Eds.), *Political Action.* London: Sage Publications.

Inglehart, R. (1981). Postmaterialism in an Environment of Insecurity. *American Political Science Review, 75*(4), 880–900.

Inglehart, R. (1990). *Culture Shift in Advanced Industrial Societies.* Princeton, NJ: Princeton University Press.

Inglehart, R. (1997). *Modernization and Postmodernization. Cultural, Economic, and Political Change in 43 Societies.* Princeton, NJ: Princeton University Press.

Inglehart, R., & Abramson, P. R. (1999). Measuring Postmaterialism. *American Political Science Review, 93*(3), 665–677.

Jamison, A. (1980). Miljökampens Historia i Skandinavien. *Natur och Samhälle, 7*(3–4), 108–128.

Jansen, A.-I. (1989). *Makt og Miljø. Om Utformingen av Natur-og Miljøvernpolitikk i Norge.* Oslo: Universitetsforlaget.

Jansen, A.-I., & Mydske, P. K. (1998). Norway: Balancing Environmental Quality and Interest in Oil. In K. Hanf & A.-I. Jansen (Eds.), *Governance and Environment in Western Europe* (pp. 181–207). London: Longman.

Jasper, J. M., & Nelkin, D. (1992). *The Animal Rights Crusade. The Growth of a Moral Protest.* New York: Free Press.

Jones, R. E., & Dunlap, R. E. (1992). The Social Bases of Environmental Concern: Have They Changed Over Time? *Rural Sociology, 57*(1), 28–47.

Jordan, G., & Maloney, W. (1997). *The Protest Business? Mobilizing Campaign Groups.* Manchester, UK: Manchester University Press.

Jørgensen, S. I. (2001). *Attac og Globaliseringen.* Oslo: Aschehoug.

Kaltenborn, B. P., Bjerke, T., & Strumse, E. (1998). Diverging Attitudes Towards Predators: Do Environmental Beliefs Play a Part? *Research in Human Ecology, 5*(2), 1–9.

Kanagy, C. L., & Nelsen, H. M. (1995). Religion and Environmental Concern: Challenging the Dominant Assumptions. *Review of Religious Research, 37*(1), 33–45.

Kanagy, C. L., & Willits, F. K. (1993). A Greening of Religion: Some Evidence from a Pennsylvania Sample. *Social Science Quarterly, 74*(3), 674–683.

Kitschelt, H., & Hellemans, S. (1990). Left–Right Semantics in the New Politics Cleavage. *Comparative Political Studies, 23*(2), 210–238.

Klausen, K. K., & Selle, P. (1995). Frivillig Organisering i Norden. In K. K. Klausen & P. Selle (Eds.), *Frivillig Organisering i Norden* (pp. 13–31). Oslo: TANO/Jurist-og Økonomiforbundets Forlag.

Klecka, W. R. (1980). *Discriminant Analysis* (Vol. 19). Beverly Hills, CA: Sage.

Klingemann, H.-D. (1995). Party Position and Voter Orientation. In H.-D. Klingemann & D. Fuchs (Eds.), *Citizens and the State* (Vol. 1, pp. 183–205). Oxford: Oxford University Press.

Knappskog, T. (1995). Grøne Parti i Finland, Noreg og Sverige. *Paper.* Department of Comparative Politics, University of Bergen.

Knutsen, O. (1995). Value Orientations, Political Conflicts and Left–Right Identification: A Comparative Study. *European Journal of Political Research, 28*(1), 63–93.

Knutsen, O. (1997). From Old Politics to New Politics. Environmentalism as a Party Cleavage. In K. Strøm & L. Svåsand (Eds.), *Challenges to Political Parties. The Case of Norway* (pp. 229–262). Ann Arbor: University of Michigan Press.

Knutsen, T., Aasetre, J., & Sagør, J. T. (1998). *Holdninger til Rovvilt i Norge.* Trondheim: Senter for Miljø og Utvikling, NTNU. (Rapport Nr. 4/98).

Kraus, S. J. (1995). Attitudes and the Prediction of Behavior: A Meta-analysis of the Empirical Literature. *Personality and Social Psychology Bulletin, 21*(1), 58–75.

Kuhnle, S., & Selle, P. (1990). Meeting Needs in a Welfare State: Relations Between Government and Voluntary Organizations in Norway. In A. Ware & R. E. Goodin (Eds.), *Needs and Welfare.* London: Sage.

Kuhnle, S., & Selle, P. (1992a). Government and Voluntary Organizations: A Relational Perspective. In S. Kuhnle & P. Selle (Eds.), *Government and Voluntary Organizations* (pp. 1–34). Aldershot, UK: Avebury.

Kuhnle, S., & Selle, P. (1992b). Governmental Understanding of Voluntary Organizations: Policy Implications of Conceptual Change in Post-war Norway. In S. Kuhnle & P. Selle (Eds.), *Government and Voluntary Organizations* (pp. 157–185). Aldershot, UK: Avebury.

Kumlin, S., & Rothstein, B. (2005). Making and Breaking Social Capital. The Impact of Welfare–State Institutions. *Comparative Political Studies, 38*(4), 339–365.

Kvaløy, S. (1972). *Økopolitikk eller EF.* Oslo: Pax.

Kvaløy, S. (1973). *Økokrise, Natur og Menneske: En Innføring i Økofilosofi og Økopolitikk.* Paper in Økologiske Fragment, nr. 4. University of Oslo.

Kvaløy Setreng, S. (1996). Framveksten av Aksjonismen og Økopolitikken. In K. Strømsnes & P. Selle (Eds.), *Miljøvernpolitikk og Miljøvernorganisering mot år 2000* (pp. 108–116). Oslo: Tano Aschehoug.

Lafferty, W., & Meadowcroft, J. (Eds.) (1996). *Democracy and the Environment: Problems and Prospects*. Cheltenham, UK: Edward Elgar Publishing.

Lafferty, W. M., & Langhelle, O. (1995). Bærekraftig Utvikling Som Begrep og Norm. In W. M. Lafferty & O. Langhelle (Eds.), *Bærekraftig Utvikling. Om Utviklingens Mål og Bærekraftens Betingelser* (pp. 13–38). Oslo: Ad Notam Gyldendal.

Lem, S. (1994). *Den tause krigen mot de fattige og mot miljøet -og hva som må gjøres*. Oslo: Forum.

Lem, S. (1996). Framtiden i Våre Hender: Punktmiljøvern Eller Samfunnsforandring? In K. Strømsnes & P. Selle (Eds.), *Miljøvernpolitikk og Miljøvernorganisering mot år 2000* (pp. 163–173). Oslo: Tano Aschehoug.

Lem, S. (2000, 24 mai). Neste bok: "Antimenneske"! *Dagbladet*.

Lewis, M. W. (1992). *Green Delusions. An Environmentalist Critique of Radical Environmentalism*. London: Duke University Press.

Lijphart, A., & Crepaz, M. M. L. (1991). Corporatism and Consensus Democracy in Eighteen Countries: Conceptual and Empirical Linkages. *British Journal of Political Research, 21*, 235–256.

Lindstrøm, U. (1997). Still Five after All These Years? In S. Lindberg & Y. Mohlin (Eds.), *Festskrift till Sten Berglund* (pp. 53–76). Vasa, Norway: Pro Facultate.

Listhaug, O. (2005). Oil wealth dissatisfaction and political trust in Norway: A resource curse?. *West Europan Politics, 28*(4), 834–851.

Listhaug, O., & Wiberg, M. (1995). Confidence in Political and Private Institutions. In H.-D. Klingemann & D. Fuchs (Eds.), *Citizens and the State* (Vol. 1, pp. 298–322). Oxford: Oxford University Press.

Lomborg, B. (1998). *Verdens Sande Tilstand*. Viby, Denmark: Centrum.

Lomborg, B. (2001). *The Skeptical Environmentalist. Measuring the Real State of the World*. London: Cambridge University Press.

Lorentzen, H. (1998). Normative Forståelser av Sivile Sammenslutninger. *Socialvetenskaplig Tidsskrift, 5*(2–3), 244–268.

Lowe, D. P., & Rüdig, W. (1986). Political Ecology and the Social Sciences: The State of the Art. *British Journal of Political Science, 16*, 513–550.

Malnes, R. (1990). *The Environment and Duties to Future Generations*. Report: Fridtjof Nansen Institute.

Martell, L. (1994). *Ecology and Society. An Introduction*. Cambridge: Polity Press.

Martinsen, S. (1996). NOAH—for dyrs Rettigheter: Dyrs Rettigheter og Miljøvern. In K. Strømsnes & P. Selle (Eds.), *Miljøvernpolitikk og Miljøvernorganisering mot år 2000* (pp. 222–227). Oslo: Tano Aschehoug.

Maslow, A. K. (1954). *Motivation and Personality*. New York: Harper & Row.

Mathews, F. (Ed.) (1996). *Ecology and Democracy*. London: Frank Cass.

McAdam, D., McCarthy, J. D., & Zald, M. N. (Eds.) (1996). *Comparative Perspectives on Social Movements. Political Opportunities, Mobilizing Structures, and Cultural Framings*. Cambridge: Cambridge University Press.

McCormick, J. (1995). *The Global Environmental Movement*. Chichester: John Wiley & Sons.

Mertig, A. G., & Dunlap, R. E. (2001). Environmentalism, New Social Movements, and the new class: a cross-national investigation. *Rural Sociology, 66*(1), 113–136.

Micheletti, M. (2003). *Political Virtues and Shopping. Individuals, Consumerism, and Collective Action*. New York: Palgrave MacMillan.

Micheletti, M., Follesdal, A., & Stolle, D. (Eds.) (2004). *Politics, Products, and Markets. Exploring Political Consumerism Past and Present.* New Brunswick, NJ: Transaction Publishers.

Milbrath, L. W. (1984). *Environmentalists. Vanguard for a New Society.* Albany, NY: SUNY Press.

Miles, L. (Ed.) (1996). *The European Union and the Nordic Countries.* London: Routledge.

Miljøstatus. (1999). http://www.mistin.dep.no/ [May 6, 1999].

Mohai, P. (1992). Men, Women, and the Environment: An Examination of the Gender Gap in Environmental Concern and Activism. *Society and Natural Resources, 5,* 1–19.

Morken, K., & Selle, P. (1994). Norway: The Women's Shelter Movement. In F. D. Perlmutter (Ed.), *Women and Social Change. Nonprofits and Social Policy* (pp. 133–157). Washington, DC: National Association of Social Workers (NASW Press).

Morris, A. D., & Mueller, C. M. (Eds.) (1992). *Frontiers in Social Movement Theory.* New Haven, CT: Yale University Press.

Müller-Rommel, F. (1985). Social Movements and the Greens. New Internal Politics in Germany. *European Journal of Political Research, 13,* 53–67.

Müller-Rommel, F. (1998). Explaining the Electoral Success of Green Parties: A Cross-National Analysis. *Environmental Politics, 7*(4), 145–154.

Næss, A. (1973). The Shallow and the Deep, Long-Range Ecology Movements; A Summary. *Inquiry, 16,* 95–100.

Nilsen, K.-E. (1996). Miljøstiftelsen Bellona: Døgnflua som Overlevde. In K. Strømsnes & P. Selle (Eds.), *Miljøvernpolitikk og Miljøvernorganisering mot år 2000* (pp. 185–198). Oslo: Tano Aschehoug.

Nisbet, R. (1962). *Power and Community.* New York: Oxford University Press.

Noe, F. P., & Snow, R. (1990). The New Environmental Paradigm and Further Scale Analysis. *Journal of Environmental Education, 21*(4), 20–26.

Nordby, T. (1994). *Korporatisme på Norsk 1920–1990.* Oslo: Universitetsforlaget.

Nunnally, J. C., & Bernstein, I. H. (1994). *Psychometric Theory* (3rd ed.). New York: McGraw-Hill.

Oelschlaeger, M. (1991). *The Idea of Wilderness. From Prehistory to the Age of Ecology.* London: Yale University Press.

Olli, E., Grendstad, G., & Wollebæk, D. (2001). Correlates of Environmental Behaviors. Bringing Back Social Context. *Environment and Behaviour, 33*(3), 181–208.

Olsen, J. P. (1983). *Organized Democracy.* Oslo: Universitetsforlaget.

Østerud, Ø. (1978). *Agrarian Structure and Peasant Politics in Scandinavia. A Comparative Study of Rural Response to Economic Change.* Oslo: Universitetsforlaget.

Østerud, Ø., Engelstad, F., & Selle, P. (2003). *Makten og Demokratiet.* Oslo: Gyldendal Akademisk.

Paehlke, R. (1989). *Environmentalism and the Future of Progressive Politics.* New Haven, CT: Yale University Press.

Paehlke, R. (1996). Environmental Challenges to Democratic Practice. In W. M. Lafferty & J. Meadowcroft (Eds.), *Democracy and the Environment. Problems and Prospects* (pp. 18–38). Cheltenham, UK: Edward Elgar Publishing.

Papakostas, A. (2004). Civilsamhallets Rationaliseringer. *Arkiv, 91,* 19–45.

Pepper, D. (1996). *Modern Environmentalism. An Introduction.* London: Routledge.

Persen, Å. B., & Ranum, N. H. (1997). *Natur og Ungdom—30 År i Veien.* Oslo: Natur og Ungdom.

Pimm, S., & Harvey, J. (2001). No Need to Worry About the Future: Environmentally, We Are Told, "Things Are Getting Better." *Nature, 414.* 149–150.

Poguntke, T. (1993). *Alternative Politics: The German Green Party*. Edinburgh: Edinburgh University Press.

Pooley, J. A., & O'Connor, M. (2000). Environmental Education and Attitudes: Emotions and Beliefs Are What Is Needed. *Environment and Behavior, 32*(5), 711–723.

Putnam, R. D. (2000). *Bowling Alone. The Collapse and Revival of American Community*. New York: Simon & Schuster.

Raaum, N. C. (1999). Kvinner i Offisiell Politikk: Historiske Utviklingslinjer. In C. Bergquist et al. (Eds.), *Likestilte Demokratier? Kjønn og Politikk i Norden*. Oslo: Universitetsforlaget.

Rayner, S., & Malone, E. (Eds.) (1998a). *Human Choice and Climate Change. Resources and Technology* (Vol. 2). Colombus, OH: Batelle Press.

Rayner, S., & Malone, E. (Eds.) (1998b). *Human Choice and Climate Change. The Societal Framework of Climate Change* (Vol. 1). Colombus, OH: Batelle Press.

Richardson, D., & Rootes, C. (Eds.). (1995). *The Green Challenge. The Development of Green Parties in Europe*. London: Routledge.

Richardson, H. (1994). *Kraftanstrengelse og Ensomhet. En Analyse av det Norske Friluftslivets Kulturelle Konstruksjoner*. Unpublished master's thesis. University of Bergen, Bergen.

Rokkan, S. (1967). Geography, Religion, and Social Class: Crosscutting Cleavages in Norwegian Politics. In S. M. Lipset & S. Rokkan (Eds.), *Party Systems and Voter Alignments: Cross-National Perspectives*. New York: The Free Press.

Rokkan, S. (1970). *Citizens, Elections, Parties*. Oslo: Universitetsforlaget.

Rommetvedt, H. (2000). Private and Public Power at the National Level. In H. Goverde, P. G. Cerny, M. Hauggard, & H. H. Lentner (Eds.), *Power in Contemporary Politics: Theories, Practices, Globalizations*. London: Sage.

Rootes, C. (Ed.) (2003). *Environmental Protest in Western Europe*. Oxford: Oxford University Press.

Rothenberg, D. (1995). Have a Friend for Lunch: Norwegian Radical Ecology Versus Tradition. In B. R. Taylor (Ed.), *Ecological Resistance Movements: The Global Emergence of Radical and Popular Environmentalism*. Albany: State University of New York Press.

Rothstein, B., & Stolle, D. (2003). Social Capital, Impartiality, and the Welfare State: An Institutional Approach. In M. Hooge & D. Stolle (Eds.), *Generating Social Capital. The Role of Voluntary Associations, Institutions and Government Policy*. New York: Palgrave Macmillan.

Rüdig, W. (Ed.) (2006). *Green Party Members in Western Europe*. Cambridge, MA: MIT Press.

Sætra, H. (1973). *Den økopolitiske sosialismen*. Oslo: Pax Forlag.

Salamon, L. M. (1987). Partners in Public Service: The Scope and Theory of Government Nonprofit Relations. In W. W. Powell (Ed.), *The Nonprofit Sector: A Research Handbook* (pp. 99–117). New Haven, CT: Yale University Press.

Salamon, L. M., & Sokolowski, W. (Eds.) (2004). *Global Civil Society. Dimensions of the Nonprofit Sector*. Bloomfield CT: Kumarian Press.

Schmitter, P. C. (1979). Still the Century of Corporatism? In P. C. Schmitter & G. Lehmbruch (Eds.), *Trends Towards Corporatist Intermediation*. London: Sage.

Seip, A. L. (1984). *Sosialhjelpsstaten Blir til. Norsk Sosialpolitikk 1740–1920*. Oslo: Gyldendal.

Sejersted, F. (1984). *Opposisjon og Posisjon: 1945–1981*. Oslo: Cappelen

Sejersted, F. (1993). *Demokratisk Kapitalisme*. Oslo: Universitetsforlaget.

Selle, P. (1996). *Frivillige organisasjonar i nye omgjevnader*. Bergen: Alma Mater.

Selle, P. (1998). Organisasjonssamfunnet—ein statsreiskap? In T. Grønlie & P. Selle (Eds.), *Ein stat? Fristillingas fire ansikt* (pp. 141–178). Oslo: Det Norske Samlaget.

Selle, P. (1999). Sivilsamfunnet tatt på alvor. In Ø. Østerud, F. Engelstad, S. Meyer, P. Selle & H. Skjeie (Eds.), *Mot en ny maktutredning* (pp. 64–93). Oslo: Ad Notam Gyldendal.

Selle, P. (2000). Norsk miljøvern er annleis. *Nytt Norsk Tidsskrift*, 16(4), 376–384.

Selle, P., & Østerud, Ø. (2006). The Eroding of Representative Democracy in Norway. *Journal of European Public Policy*, *13*(4).

Selle, P., & Øymyr, B. (1995). *Frivillig Organisering og Demokrati*. Oslo: Det Norske Samlaget.

Selle, P., & Strømsnes, K. (1996). Norske Miljøvernorganisasjoner: En Demokratisk Folkebevegelse? In K. Strømsnes & P. Selle (Eds.), *Miljøvernpolitikk og Miljøvernorganisering mot år 2000* (pp. 261–291). Oslo: Tano Aschehoug.

Selle, P., & Strømsnes, K. (1998). Organised Environmentalists: Democracy as a Key Value? *Voluntas*, *9*(4), 319–343.

Selle, P., & Strømsnes, K. (2001). Membership and Democracy. In P. Dekker & E. M. Uslaner (Eds.), *Social Capital and Politics in Everyday Life*. London: Routledge.

Siaroff, A. (1999). Corporatism in 24 Industrial Democracies: Meaning and Measurement. *European Journal of Political Research*, *36*(2), 175–205.

Sivesind, K. H., Lorentzen, H., Selle, P., & Wollebæk, D. (2002). *The Voluntary Sector in Norway. Composition, Changes and Causes*. Oslo: Institutt for samfunnsforskning.

Skjeie, H. (1992). *Den Politiske Betydningen av Kjønn. En studie av Norsk Topp-Politikk*. Oslo: Institutt for Samfunnsforskning. (Rapport 92:11).

Sklar, K. K. (1993). The Historical Foundations of Women's Power in the Creation of the American Welfare State (1830–1930). In S. Koven & S. Michel (Eds.), *Mothers of a New World. Maternalist Politics and the Origins of Welfare States*. New York: Routledge.

Skocpol, T. (1979). *States and Social Revolutions: A Comparative Analysis of France, Russia and China*. Cambridge: Cambridge University Press.

Skocpol, T. (1992). *Protecting Soldiers and Mothers*. Cambridge, MA.: Belknap Press of Harvard University Press.

Skocpol, T. (2003). *Diminished Democracy. From Membership to Management in American Civil Life*. Norman: University of Oklahoma Press.

SSB. (1975). *Population in Municipalities 1974–1975*. Oslo: Statistisk Sentralbyrå.

SSB. (1985). *Population in Municipalities 1983–1985*. Oslo: Statistisk Sentralbyrå.

SSB. (1995). *Population Statistics 1995. Volume I. Population Changes in Municipalities 1993–1995*. Oslo-Kongsvinger: Statistisk Sentralbyrå.

Stern, P. C. (2000). Toward a Coherent Theory of Environmentally Significant Behavior. *Journal of Social Issues,* 56(3), 407–424.

Stolle, D., & Hooge, M. (2004). Inaccurate, Exceptional, One-Sided or Irrelevant? The Debate About the Alleged Decline of Social Capital and Civic Engagement in Western Societies. *British Journal of Political Science*, 35, 149–167.

Strandberg, U. (2006). Introduction: Historical and Theoretical Perspectives on Scandinavian Political Systems. *Journal of European Public Policy*, *13*(4).

Strømsnes, K. (2001). *Demokrati i Bevegelse*. (R0111) Bergen: LOS-senteret.

Strømsnes, K. (2003). *Folkets Makt*. Oslo: Gyldendal Akademisk.

Strømsnes, K., Grendstad, G., & Selle, P. (1996). *Miljøvernundersøkelsen 1995*. Dokumentasjonsrapport (9616). Bergen: LOS-senteret.

Strømsnes, K., & Selle, P. (Eds.) (1996). *Miljøvernpolitikk og Miljøvernorganisering mot år 2000*. Oslo: Tano Aschehoug.

Søgård, C. (1997). Fra Rebeller til Konsulenter. En Casestudie av Miljøvernorganisasjonen Bellona. In A. Nilsen (Ed.), *Miljøsosiologi. Samfunn, Miljø og Natur* (pp. 93–106). Oslo: Pax Forlag.

Sørensen, Ø. (Ed.) (1998). *Jakten på det Norske: Perspektiver på Utviklingen av en Norsk Nasjonal Identitet på 1800-Tallet*. Oslo: Ad Notam Gyldendal.

Terragni, L., & Kjærnes, U. (2005). Ethical Consumption in Norway: Why Is It So Low? In M. Boström, A. Føllesdal, M. Klintman, M. Micheletti, & M. P. Sørensen (Eds.), *Political Consumerism: Its Motivations, Power, and Conditions in the Nordic Countries and Elsewhere. Proceedings from the 2nd International Seminar on Political Consumerism, August 26–29, 2004*. (pp. 471–485). Oslo: TemaNord.

Thompson, M. (1992). The Dynamics of Cultural Theory and Their Implications for the Enterprise Culture. In S. Hargreaves Heap & A. Ross (Eds.), *Understanding the Enterprise Culture. Themes in the Work of Mary Douglas* (pp. 182–202). Edinburgh: Edinburgh University Press.

Thompson, M. (2000). Global Networks and Local Cultures. What Are the Mismatches and What Can Be Done About Them? In C. Engel & K. H. Keller (Eds.), *Understanding the Impact of Global Networks on Local, Social, Political and Cultural Values* (Vol. 42, pp. 113–129). Baden-Baden: Nomos Verlagsgesellschaft.

Thompson, M., Ellis, R., & Wildavsky, A. (1990). *Cultural Theory*. Boulder, CO: Westview Press.

Thompson, M., Grendstad, G., & Selle, P. (Eds.) (1999). *Cultural Theory as Political Science*. London: Routledge.

Tindall, D. B., Davies, S., & Mauboules, C. (2003). Activism and Conservation Behavior in an Environmental Movement: The Contradictory Effects of Gender. *Society & Natural Resources, 16*(10), 909–932.

Tjernshaugen, A. (1999). På Jakt Etter Penger som Ikke Lukter. *Bellona Magasin, 11*(2), 38–41.

Togeby, L. (1984). *Politik: Også en Kvindesag. En Sammenlignende Undersøgelse af Unge Kvinders og Mænds Politiske Deltagelse*. Århus: Politica.

Tourain, A. (1987). Social Movements: Participation and Protest. *Scandinavian Political Studies, 10*(3), 207–222.

Tranvik, T., & Selle, P. (2003). *Farvel til Folkestyret? Nasjonalstaten og de nye Nettverkene*. Oslo: Gyldendal Akademisk.

Tranvik, T., & Selle, P. (2005). Transformation of the Democratic Infrastructure in Norway. The Organized Society and State-Municipal Relations. *West European Politics, 28*(4), 852–871.

Tranvik, T., Thompson, M., & Selle, P. (2000). Doing Technology (and Democracy) the Pack-Donkey's Way. The Technomorphic Approach to ICT Policy. In C. Engel & K. H. Keller (Eds.), *Governance of Global Networks in the Light of Differing Local Values* (Vol. 43, pp. 155–195). Baden-Baden: Nomos Verlagsgesellschaft.

Uslander, E. M. (2002). *The Moral Foundation of Trust*. New York: Cambridge University Press.

Valen, H. (1992). *Valg og Politikk*. Oslo: NKS-Forlaget.

van der Heijden, H.-A. (1997). Political Opportunity Structure and the Institutionalisation of the Environmental Movement. *Environmental Politics, 4*, 25–50.

Van Liere, K. D., & Dunlap, R. E. (1980). The Social Bases of Environmental Concern: A Review of Hypothesis, Explanations and Empirical Evidence. *Public Opinion Quarterly, 44*, 181–199.

Warren, K. J. (Ed.) (1994). *Ecological Feminism*. London: Routledge.

White, L. (1967). The Historical Roots of Our Ecological Crisis. *Science, 155*, 1203–1207.

Wildavsky, A. (1987). Choosing Preferences by Constructing Institutions: A Cultural Theory of Preference Formation. *American Political Science Review, 81*(1), 3–21.

Wildavsky, A. (1995). *But Is It True? A Citizen's Guide to Environmental Health and Safety Issues.* Cambridge, MA: Harvard University Press.

Williamson, P. J. (1989). *Corporatism in Perspective.* London: Sage.

Witoszek, N. (1997). The Anti-Romantic Romantics; Nature, Knowledge, and Identity in Nineteenth-Century Norway. In M. Teich, R. Porter & B. Gustafsson (Eds.), *Nature and Society in Historical Context* (pp. 209–227). Cambridge: Cambridge University Press.

Wolkomir, M., Futreal, M., Woodrum, E., & Hoban, T. (1997). Substantive Religious Belief and Environmentalism. *Social Science Quarterly, 78*(1), 96–108.

Wollebæk, D. (2000). *Participation in Voluntary Associations and the Formation of Social Capital* (R0006). Bergen: LOS-senteret.

Wollebæk, D., & Selle, P. (2002a). *Det nye Organisasjonssamfunnet.* Bergen: Fagbokforlaget.

Wollebæk, D., & Selle, P. (2002b). Does Participation in Voluntary Organizations Contribute to Social Capital? The Importance of Intensity, Scope, and Type. *Nonprofit and Voluntary Sector Quarterly, 31*(2), 32–61.

Wollebæk, D., & Selle, P. (2002c). Passive Support. No Support at All? *Nonprofit Management & Leadership, 13*(2), 187–203.

Wollebæk, D., & Selle, P. (2003). Generations and Organizational Change. In P. Dekker & L. Halman (Eds.), *The Values of Volunteering. Cross-Cultural Perspectives.* New York: Kluwer Academic/Plenum.

Wollebæk, D., Selle, P., & Lorentzen, H. (2000). *Frivillig Innsats.* Bergen: Fagbokforlaget.

Wörlund, I. (2005). Miljöpartier i Sverige och Norge. In M. Demker & L. Svåsand (Eds.), *Svensk–Norsk Partiutveckling 1905–2005.* Stockholm: Santérus Förlag.

WWF. (2000). Opprettelse av Medlemsforening for WWF i Norge. *Verdens Natur, 15*(2), 12–14.

Wyller, T. C. (1999). *Demokratiet og Miljøkrisen. En Problemskisse.* Oslo: Universitetsforlaget.

Young, S. C. (1992). The Different Dimensions of Green Politics. *Environmental Politics, 1*(1), 9–44.

Acknowledgments

We are indebted to a number of colleagues for comments and advice and to several institutions for funding both covering the research process from data collection to preliminary versions of various chapters. We have had the opportunity to present and discuss our research and analyses at a number of seminars, workshops, and conferences over the years. We have specifically benefited from the following venues: International Institute for Applied Systems Analysis, Laxenburg, Austria, 7–9 August 1995; ECPR Joint Sessions of Workshops at Oslo, 29 March–3 April 1996, at Berne, 26 February–4 March 1997, and at Copenhagen, 14–20 April 2000; the Second and Third International Conference of the International Society for Third Sector Research (ISTR) in Mexico City, 18–21 July 1996 and at Geneva, 8–11 July 1998. We are grateful to the participants of these conferences for comments. We would also like to thank the following persons for comments and critical observations at various stages during this project: Helmut Anheier, William A. Maloney, Arne Næss, Eero Olli, Sanjeev Prakash, Sigmund Kvaløy Setreng, Karl Henrik Sivesind, Tommy Tranvik, Dag Wollebæk, Øyvind Østerud, and Bernt Aardal.

Our project received financial support from the Research Council of Norway under the Environment and Power research program, the Norwegian Research Centre in Organization and Management (later renamed the Stein Rokkan Centre) at the University of Bergen, the Department of Comparative Politics and School of Social Sciences both at University of Bergen. We are grateful to these institutions. Studies of environmentalism and the environmental movement was also an important part of the Norwegian Public Audit on Power and Democracy (1998–2003). This audit was a comprehensive political science project initiated and funded by the Norwegian government (Østerud, Engelstad, & Selle, 2003). Indeed, it is a testament to the state-friendly-society hypothesis that we discuss in this book that such public audits are found in Norway and a few neighboring countries only.

Appendix A:
Sample Weights

Survey of Environmentalism 1995 consists of 1 sample drawn from the general Norwegian population and 12 samples drawn from 12 environmental organizations. Technically, the 12 organizational populations are subpopulations of the general Norwegian population. The main distinction therefore runs between the general population sample and the 12 organizational samples. Our primary research interest concurs with this distinction. The other research interest is the difference among the four types of environmental organization.

The initial selection of the 12 organizations, from each of which 1 sample was drawn, was carried out with a view to represent the environmental field in the best possible way (e.g., year of foundation, organizational style, as well as environmental and ecological agenda). Including any other organization in our sample would not, we argue, add anything significantly to our understanding of our unique and anomalous case. Consequently, we treat the 12 organizations as the population of environmental organizations. It follows that each sample should be of equal weight. We therefore developed weights for respondents in each of the 12 samples so that each organization counts as 1 in the analysis (see Table A). Weights are used in all analyses in which organizations are units of observation.

TABLE A. Sample weights.

	Net sample	Ideal frequency	Weight
The Norwegian Society for the Conservation of Nature	306	174	0.56863
World Wide Fund for Nature	180	174	0.96667
Nature and Youth	183	174	0.95082
The Future in Our Hands	210	174	0.82857
The Bellona Foundation	118	174	1.47458
Greenpeace Norway	188	174	0.92553
Green Warriors of Norway	128	174	1.35938
The Norwegian Mountain Touring Association	146	174	1.19178
Norwegian Organization for Ecological Agriculture	204	174	0.85294
NOAH—for animals rights	224	174	0.77679
Women–Environment–Development	87	174	2.00000
The Environmental Home Guard	114	174	1.52632
	2088	2088	
General population	1023	1023	1.00000

Appendix B:
Internet Addresses

Adapting to technological demands, almost all of the organizations included in this study have developed homepages on the Internet. The following are the addresses as of May 2006:

Green Warriors of Norway	http://www.miljovernforbundet.no/
Greenpeace Norway	http://www.greenpeace.org/norway/
Nature and Youth	http://www.nu.no/
NOAH—for animal rights	http://www.dyrsrettigheter.no/
The Bellona Foundation	http://www.bellona.no
The Environmental Home Guard	http://www.gronnhverdag.no/
The Future in Our Hands	http://www.framtiden.no/
The Norwegian Mountain Touring Association	http://www.turistforeningen.no/
The Norwegian Organization for Ecological Agriculture	http://www.oikos.no/
The Norwegian Society for the Conservation of Nature	http://www.naturvern.no/
Women–Environment–Development	(Organization folded)
World Wide Fund for Nature	http://www.wwf.no/core/index.asp

Index